Earthquakes and the Urban Environment

Volume II

Author

G. Lennis Berlin
Associate Professor
Department of Geography
Northern Arizona University
Flagstaff, Arizona

CRC Press is an imprint of the
Taylor & Francis Group, an **informa** business

First published 1980 by CRC Press
Taylor & Francis Group
6000 Broken Sound Parkway NW, Suite 300
Boca Raton, FL 33487-2742

Reissued 2018 by CRC Press

Library of Congress Cataloging in Publication Data

Berlin, Graydon Lennis, 1943-
Earthquakes and the urban environment.

Bibliography: p.
Includes index.
1. Earthquakes. 2. Earthquakes and building.
I. Title.
QE539.848 551.2'2 77-16131
ISBN 0-8493-5174-1

A Library of Congress record exists under LC control number: 77016131

Publisher's Note
The publisher has gone to great lengths to ensure the quality of this reprint but points out that some imperfections in the original copies may be apparent.

Disclaimer
The publisher has made every effort to trace copyright holders and welcomes correspondence from those they have been unable to contact.

ISBN 13: 978-1-315-89245-0 (hbk)
ISBN 13: 978-1-351-07155-0 (ebk)

Visit the Taylor & Francis Web site at http://www.taylorandfrancis.com and the
CRC Press Web site at http://www.crcpress.com

ACKNOWLEDGMENTS

I would like to express my deepest gratitude to the many people who contributed substantially to the creation of this book. The entire manuscript was reviewed by James R. Underwood, Jr., Chairman-Department of Geology, Kansas State University and John Kelleher, Seismologist, U.S. Nuclear Regulatory Commission. Their authoritative criticisms and suggestions are largely responsible for any success of this book. Because it was not possible to make all the changes suggested, the author is solely at fault for any errors or omissions that may remain. The following individuals reviewed portions of the manuscript related to their research specialities: Jack Barrish (Jack Barrish Consulting Engineers), James H. Dieterich (U.S. Geological Survey), Ajit S. Virdee (California State University at Sacramento), James H. Whitcomb (California Institute of Technology), and Peter I. Yanev (URS/John A. Blume & Associates, Engineers). The many individuals who provided photographs vital to the completion of this book are acknowledged in the figure captions.

Two colleagues at Northern Arizona University deserve special thanks. Dominic J. Pitrone provided countless constructive suggestions and invaluable data collection assistance, and Howard G. Salisbury, III, Chairman-Department of Geography and Planning, was untiring in his efforts to accommodate my many special requests. Cartographic assistance was provided by Michael Schramm and Nat Garcia. Carolyn Waller and Virginia Hall typed the entire manuscript.

The talented editors at CRC Press, namely Sandy Pearlman, Jeffrey Eldridge, Terri Weintraub, Beth Frailey, Barbara Perris, and Gayle Tavens, contributed a great deal to the refining of much of the text. These individuals also were most understanding concerning the many delays caused by the author.

My grateful thanks go to my wife Judy for her encouragement and forbearance over the several years that it took to complete this project.

Graydon Lennis Berlin
Flagstaff, Arizona
November 1978

THE AUTHOR

Graydon Lennis Berlin was born in St. Petersburg, Pennsylvania, on May 21, 1943, and educated in the public schools there. He received a B.S. degree in 1965 from Clarion State College (Earth-Space Science), an M.A. degree in 1967 from Arizona State University (Geography), and a Ph.D. degree in 1970 from the University of Tennessee (Geography). He began his educational career in 1968 as an Assistant Professor of Geography and Research Associate at Florida Atlantic University. He joined the faculty at Northern Arizona University in 1969, attaining the rank of Associate Professor of Geography in 1975. Between 1969 and 1978, Dr. Berlin was also a Research Geographer on a part time basis with the U.S. Geological Survey. At the present time he is the Director of the Advanced Training of Foreign Participants in Remote Sensing Program, a joint venture of Northern Arizona University and the U.S. Geological Survey.

Dr. Berlin is a member of several national professional organizations and was elected Chairman of the Geography Division of the Arizona Academy of Science in 1974. He is also a member of Gamma Theta Upsilon, the National Professional Geographic Fraternity and a Full Member of Sigma Xi, the Scientific Research Society of North America. Dr. Berlin is a biographee in American Men and Women of Science, Who's Who in the West, and the Dictionary of International Biography. He is the author of more than 30 journal articles and government reports.

DEDICATION

To Judy, Jodi, Mom, and Dad

TABLE OF CONTENTS

TABLE OF CONTENTS

Chapter 1

EARTHQUAKE PREDICTION

The scientists first concluded in 1970 from anomalies in the earthquake pattern that an earthquake might be coming. In June 1974, observations of further changes in the earthquake pattern, tilting of the land surface, changes in water level in wells, changes in electric current in the ground, and strange animal behavior confirmed this conclusion. More seismographs and tiltmeters were moved into the area. On December 20, 1974, local government was warned to expect a large earthquake soon, and, in mid-January 1975, warning was given that the quake was imminent. On January 28, villages were warned to be prepared. Extra seismographs were set up.

Observations in the threatened area continued until February 1, when indications of an impending earthquake began to mount. A minor tremor was detected in an area that had not recently experienced one. The next day, there were seven more. On February 3, the minor tremors increased further, and more shocks were felt.

These events led the scientists to call an emergency conference at 7 p.m. on February 3 to report to authorities their prediction that a strong earthquake would probably occur in the very near future. By the afternoon of February 4, the seismic activity had leveled off, but this was judged to be the calm before the storm. At 2 p.m., people were told to expect a major quake within 2 days. Shops were shut, and general evacuation of buildings was ordered in two counties. At 6 p.m. that night in one village, the people were warned, "A strong earthquake will probably occur tonight. We insist that all people leave their homes and all animals leave their stables. The people from the cinema team will show four feature films outside for us tonight."

One and one-half hours later, the earthquake, measured at 7.3 on the Richter scale, struck.[631]

As noted by Hamilton,[631] this passage was not extracted from the writings of science fiction; rather, it summarizes the course of events that reportedly preceded the February 4, 1975 earthquake that struck the Liaoning Province in the People's Republic of China. Because of the accuracy of the prediction, more than one million people were evacuated from their homes, an action that probably saved tens of thousands of lives.[632] The prediction emanated from a program that was less than 10 years old.

Most earth scientists believe that similar scenarios will become increasingly more common. In addition to the People's Republic of China, where several destructive earthquakes have been successfully predicted in the last 5 years, the development of a reliable earthquake forecasting capability is also a national goal in Japan, the Soviet Union, and the U.S. — countries where scientists have predicted several small seismic events.[633]

The prediction of shallow-focus earthquakes on a routine and reliable basis is, without question, one of the great challenges of science. However, significant strides towards the attainment of this goal have been realized in just the past few years. During this period, it has been established that a number of earthquakes were preceded by certain *geophysical anomalies* in their source regions[633] that had been predicted earlier from laboratory and theoretical studies. These anomalies are also called *precursors* or *premonitory phenomena.* The ability to detect, measure, and assess precursors will hopefully lead to predictions in their truest sense — that is, accurate and consistent specifications of a pending earthquake's location, time of occurrence, and size.

Several distinct models have been developed to explain the formation of earthquake precursors. The *dilatancy mechanism* of rock mechanics, *premonitory fault creep,* and a *propagating wave front* are key components in individual models. Dilatancy, as operative in laboratory studies, defines an inelastic volume increase in a rock that is undergoing deformation; the expanded volume is caused by the opening of microcracks in the specimen before it ruptures. In the fault creep model, two phases of premonitory fault creep prepare a fault for a seismic-slip event. A propagating wave front defines a *moving stress force* of a probable deep-seated origin that produces rapid regional deformation.

This chapter is concerned with an analysis of (1) high-priority precursor regions, (2) promising earthquake precursors, (3) various earthquake precursor models, and (4) prediction programs in Japan, the Soviet Union, the People's Republic of China, and the U.S. The social implications of earthquake predictions are discussed in Volume III, Chapter 2.

I. HIGH-PRIORITY PRECURSOR REGIONS

The components of plate tectonics can be used as a model for making generalized predictions. For example, earthquakes are much more apt to occur along plate boundaries than in plate interiors, and magnitudes are smaller for divergent plate boundary shocks than for those centered along transform and convergent boundaries. However, it may be possible to improve the geographic and magnitude specificity (especially the former) for large and great events by locating *seismic gaps* or temporary quiescent areas within active segments of plate boundaries. The seismic gap technique has also been used to predict potential minor and moderate earthquakes along relatively short fault segments.

Because seismic gaps identify potential high-risk areas, they can serve as high-priority locales for deploying dense arrays of instrumentation in the search for precursors that may predate small, moderate, large, and/or great earthquakes. This strategy is now being applied to the gap along the San Andreas fault which last ruptured in 1857 (Ft. Tejon earthquake).

Attempts have been made to automatically identify future earthquake sites by computer analysis of various geologic and seismologic parameters. This technique is termed *pattern recognition.* Suspected areas of high seismic risk, defined by pattern recognition, can also serve as sites for precursor searches.

A. Seismic Gaps for Large and Great Earthquakes

Page[634] offers the following explanation of the seismic gap principle:

> If there is relative motion between two plates at one point on their common boundary, then over a sufficiently long interval of time — a century or more — movement can be expected at every point on their boundary. Seismic gaps along plate margins are thus viewed as temporary features indicative of areas where elastic strain has been accumulating without release in earthquakes. The oldest seismic gaps are considered to be the likeliest sites for future large earthquakes.

Gaps are usually delineated by plotting the *rupture zones* of large earthquakes rather than by plotting epicenters which express only the points of initial rupture. Because it is often difficult to map ruptures directly (many are in submarine areas and others might not show breaks at the surface), the distribution of aftershocks is used to infer rupture lengths.[635]

Fedotov,[636] one of the first to use the seismic gap technique, plotted the rupture zones of large, near-surface earthquakes along the Japan-Kurile-Kamchatka arc. He identified several gaps where there had been no ruptures for many years and concluded that they were likely sites for large earthquakes in the future. Kelleher et al.[637] report that since Fedotov's 1965 predictions, three large earthquakes ($M_s \geq 7.0$) have filled gaps delineated by Fedotov.

Similar to the procedure used by Fedotov, Allen et al.[638] constructed a strain-release map of southern California for the period from 1934 to 1963 and identified several aseismic areas that they thought were likely sites for large earthquakes along the San Andreas fault. In addition, Tobin and Sykes[639] proposed that two zones along the seismic belt of the northeast Pacific Ocean were likely sites for future shocks because the areas had been essentially aseismic for many years.

Several investigators have identified seismic gaps in and near Japan.[640-643] To date, the sites of the August 11, 1969 Hokkaido-Toho-Oki (M_s = 7.8) and June 17, 1973 Nemuro-Oki (M_s = 7.7) earthquakes were successfully predicted by Mogi[640] and Utsu,[642] respectively. The gap struck by the 1973 earthquake had been designated an "area of special observation" (i.e., an area to monitor for short-term precursors) by the Japanese Government's Coordinating Committee for Earthquake Prediction (CCEP) in 1970.[644]

Sykes[645] relocated all aftershocks from M_s >7.0 earthquakes from the Aleutian Islands to offshore British Columbia from 1930 to 1970 to delineate rupture zones for each earthquake. Upon completion, it was observed that the plate boundary had been ruptured by large shocks except for three segments which Sykes concluded were likely sites for future earthquakes: (1) the western Aleutians — Commander Islands, (2) southern Alaska near a sequence of large earthquakes in 1899 and 1900, and (3) southeast Alaska. Page[634] reported that an M_s = 7.6 earthquake occurred near the community of Sitka (area #3) on July 30, 1972. The rupture was centered along a segment of the Fairweather fault that separates the American and Pacific plates. Kelleher and Savino[646] supported Sykes' analysis by noting that the Sitka region, although having moderate earthquakes in the mid-1960s, became extremely aseismic as the time of the main shock approached. Sykes[635] notes that the region of the great 1964 Alaskan earthquake had been inactive from at least 1900 to 1964.

A comprehensive study concerning potential sites for large earthquakes in the near future (i.e., 10 or a few tens of years) as determined by seismic gaps has been completed by Kelleher et al.[637] They studied parts of the Pacific and Caribbean plate margins (Figure 1) and determined two types of potential earthquake sites: (1) those having satisfied initial criteria — part of a major, shallow seismic belt dominated by strike-slip or thrust faulting with no rupturing for at least 30 years,and (2) those meeting initial criteria plus at least one supplemental criterion — a historical record of one or more large earthquakes occurring in a segment, historical data suggesting that a recurrence interval is near, or that the segment appears to be the next site for a migratory earthquake sequence progressing regularly in time and space (Figure 2).

The authors stress that Figure 2 should be regarded only as a most general type of prediction map. Its specific value lies in the fact that certain of the segments possess *special seismic potential*. These should be instrumented with a variety of seismological, geodetic, and geophysical sensors for analyzing possible precursors that might provide data for the accurate prediction of large earthquakes.[637]

In reference to the San Andreas fault, some scientists believe that creep and small- to moderate-sized earthquakes relieve an adequate amount of accumulating strain to prevent major earthquakes from occurring along those segments experiencing such activity. For example, Allen[647] divided this fault into five segments — three unlocked (active) and two locked (inactive). The two inactive zones coincide with the rupture zones of the January 9, 1857 Ft. Tejon and April 18, 1906 San Francisco earthquakes. Allen believes infrequent but great earthquakes will occur here in the future because strain continues to accumulate.

Kelleher et al.[637] contend that strain along a plate boundary is relieved primarily by periodic large earthquakes and not by creep or small shocks. They argue that areas experiencing creep should not be totally excluded as potential sites for large earthquakes. Part of the rationale supporting this view came from the laboratory studies of Scholz et al.[648] They discovered that stick-slip was always preceded by a small amount of creep or stable frictional sliding in granite specimens subjected to compres-

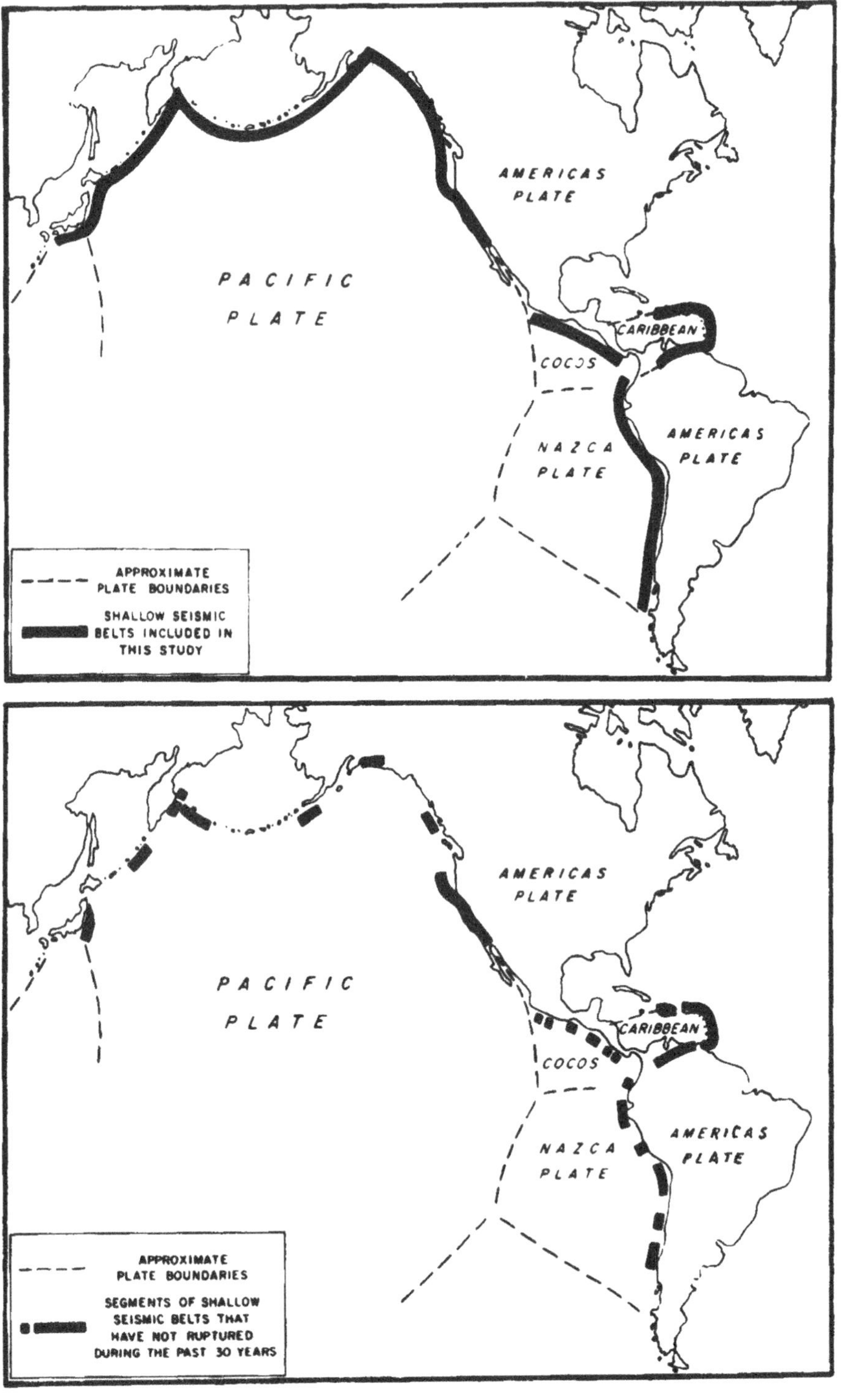

FIGURE 1. Major seismic belts examined (top) and seismic segments that have not ruptured during the past 30 years. (From Kelleher, J., Sykes, L., and Oliver, J., *J. Geophys. Res.*, 78, 2551, 1973. Copyrighted by American Geophysical Union. With permission.)

sional stress (discussed later in this chapter). This would be indicative of high, not low,

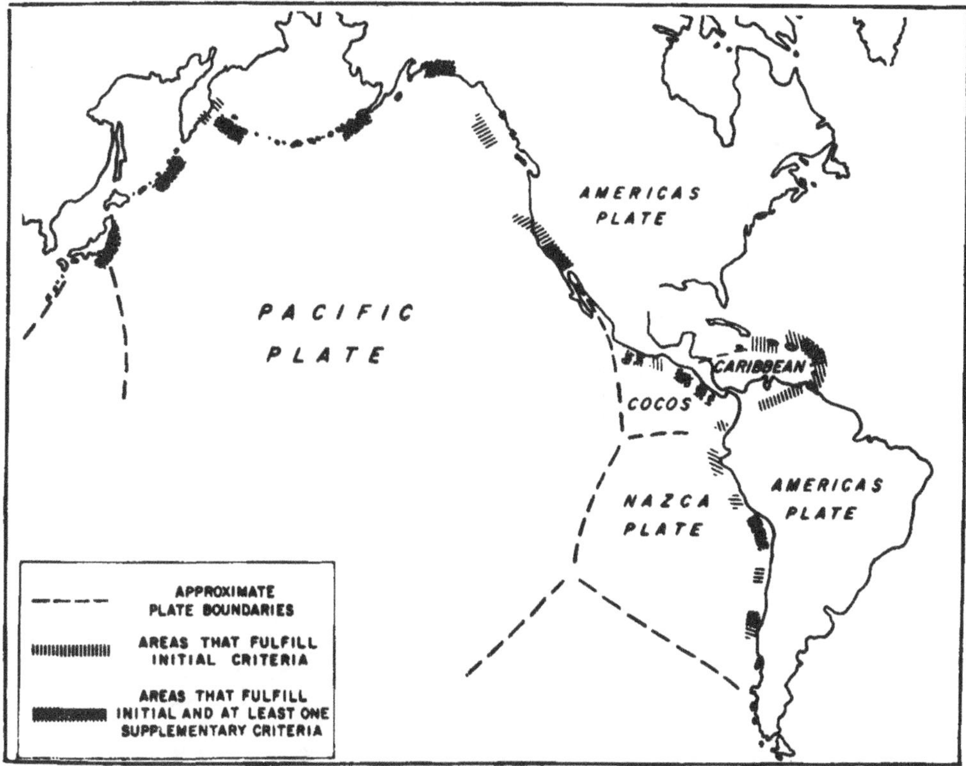

FIGURE 2. Likely locations for large earthquakes along segments of Pacific and Caribbean plate boundaries that fulfill initial or initial and supplementary criteria. See text for criteria definitions. (From Kelleher, J., Sykes, L., and Oliver, J., *J. Geophys. Res.*, 78, 2548, 1973. Copyrighted by American Geophysical Union. With permission.)

stress. Therefore, Kelleher et al. maintain that until clear evidence to the contrary is forthcoming large earthquakes should be anticipated along virtually all of the San Andreas fault (Figure 3).

B. Linear Migration of Large Earthquakes

Several investigators have reported on large shallow-focus earthquakes following a linear (sequential) migration along a fault zone. For example, Kelleher[649] and Sykes[645] note that five out of six large earthquakes occurring along the Aleutian arc (146° to 171°E) since 1938 progressed in space and time from east to west. Based upon this space-time trend, Kelleher[649] predicted a large earthquake at approximately 56°N, 158°W for sometime between 1974 and 1980. This area was struck by a large earthquake in 1938.

Kelleher[650] also discovered a north to south migration pattern for large earthquakes along much of the Chilean seismic belt. Subsequent to submitting his article for publication, a M_s = 7.6 earthquake occurred on July 9, 1971. Although the magnitude was smaller than expected, the event fits this predicted north to south trend.

Anderson[651] has proposed that the linear migration of larger earthquakes along a convergent plant boundary (e.g., Aleutians) might be caused by great *decoupling earthquakes* (i.e., a trench event in which the boundary separating the underthrusting plate and restraining plate is broken, resulting in a decoupling of the two converging plates). A decoupling event is thought to cause increasing stresses along adjacent arc segments

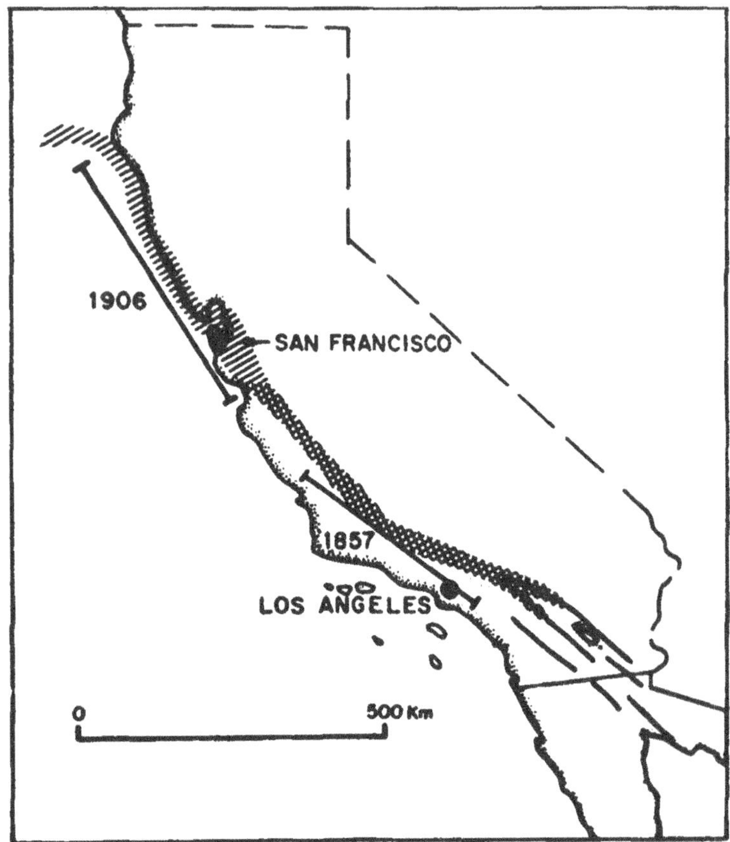

FIGURE 3. Segments of the San Andreas fault system fulfilling initial or initial and supplemental criteria. See text for criteria definitions. Line segments mark the approximate rupture zones of the January 9, 1857 Ft. Tejon and April 18, 1906 San Francisco earthquakes. (From Kelleher, J., Sykes, L., and Oliver, J., *J. Geophys. Res.*, 78, 2578, 1973. Copyrighted by American Geophysical Union. With permission.)

due to increased plate motions in the vicinity of the decoupling earthquake as well as stress wave diffusion from the event (i.e., a stress drop which diffuses in all directions, but especially along the plate boundary).

A progressional trend has also been discovered along the North Anatolian strike-slip fault in central Turkey.[652-655] Dewey[655] reports that the seven largest shocks (M_s = 6.8 to 8.0) occurring along the fault from 1939 through 1967 displayed a linear migration from east to west (Figure 4). These seven earthquakes ruptured the fault for an aggregate distance of approximately 800 km.

Savage[656] believes that the linear pattern is explainable by a kinematic-wave model. In this model, a creep wave is created by an earthquake releasing an avalanche of dislocations. The wave subsequently moves down the fault in the direction of dislocation flow until it strikes a locked section of the fault. The dislocations accumulate there, increasing the local stresses. If there are a sufficient number of dislocations in the wave, the stresses will increase to a level causing slip, and an earthquake, at the locked section. This earthquake gives rise to a new avalanche of dislocations. In the case of the right-lateral North Anatolian fault, the dislocations would migrate to the west.

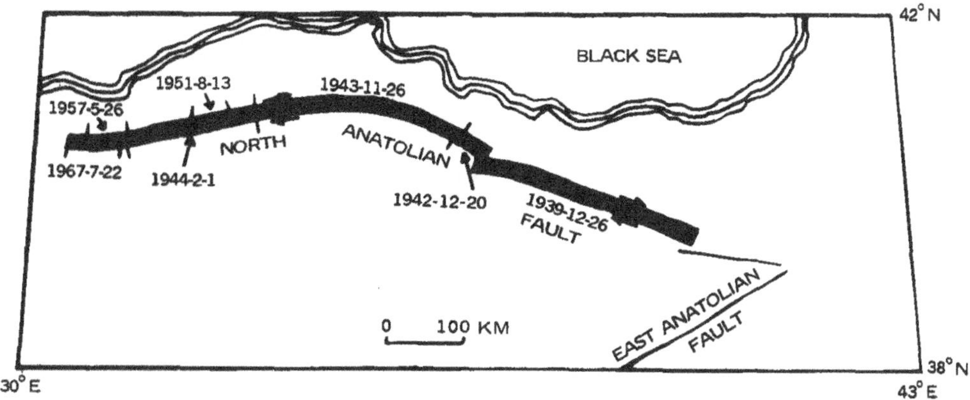

FIGURE 4. *Surface faulting on the North Anatolian fault in central Turkey between 1939 and 1967. Note the temporal migration to the west. (From Dewey, J. W., Earthquake Inf. Bull., 6, 13, 1974.)*

Nikonov's[657,658] spatial and temporal analysis of $M_S \geq 6.5$ earthquakes along the Gissar-Kokshaal and the Hindu-Kush-Darvaz-Karakul thrust-type fault zones in Soviet Central Asia indicates a progression from their flanks towards the center of the Parmir arcs. The rate of migration varies from 1 to 2 to 3 to 6 km/year. No systematic migrational pattern is discernible for $M < 6.5$ earthquakes. By using the rates and directions of migration, Nikonov has delineated possible sites for large earthquakes before the end of the century.

Nikonov[657] states that the main fault zones in Soviet Central Asia are controlled by a regional compressive system with the dominant principal stress oriented north-south. The lack of a pattern for smaller shocks would be caused "by the stress distribution in limited areas, and therefore . . . not directly governed by regional patterns."

C. Seismic Gaps for Minor and Moderate Earthquake Predictions

The seismic gap technique has been used in California to predict potential earthquakes of moderate and minor magnitudes. Like gaps that may be future sites of large or great earthquakes, these seismic gaps could also serve to locate high-priority sites to search for potential precursors.

Ellsworth and Wesson[659] analyzed a 21-km segment of the central San Andreas fault between Melendy Ranch and Cienega School where four moderate earthquakes (M_L's = 5.0, 4.7, 4.0, and 4.0) occurred between December 1971 and January 1973. It was discovered that (1) slip surfaces (determined by aftershock distributions) for earthquake pairs abuted each other with a slight overlap at both ends of the 21-km segment and (2) a 4-km-long gap existed between the two composite slip zones (Figure 5). Based upon the hypothesis that clusters of small shocks occurring in the vicinity of a main event hypocenter are symptomatic of conditions favorable for the initiations of rupture (small tremors had preceded the above four quakes in the immediate vicinity of their hypocenters), they concluded that a $M_L = 4.5$ earthquake would fill the gap within several months after April 1973 (Figure 5). The magnitude estimate was based upon the length of rupture needed to fill the gap. No single earthquake occurred, but the prediction was a milestone because it represented the first prediction made by scientists of the U.S. Geological Survey (USGS). The strain in the gap was subsequently released by several small-magnitude shocks and perhaps by creep.

Thatcher et al.[660] recently reported on two gaps along the San Jacinto fault (part of the San Andreas system) in southern California (Figure 5 in Volume I, Chapter 2); significant right-lateral slip has not occurred in either gap since 1890. One gap runs

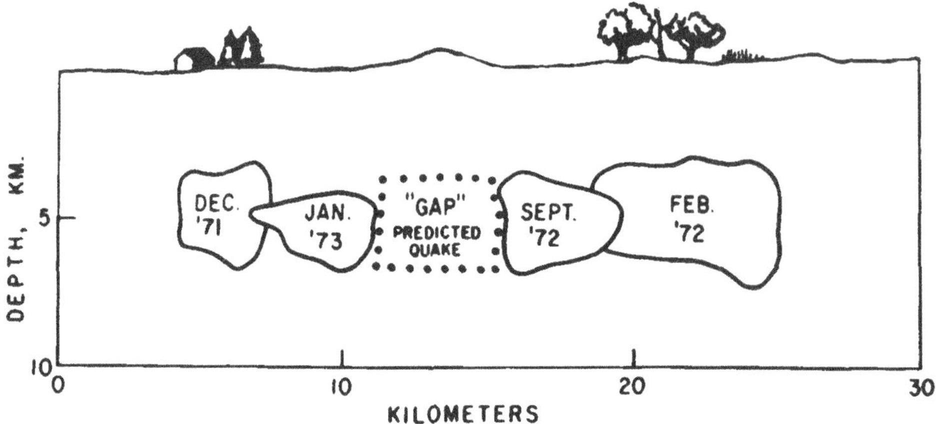

FIGURE 5. A 4-km *seismic gap* identified by Ellsworth and Wesson[655] as a site for a future earthquake of moderate size along the San Andreas fault between Melendy Ranch and Cienega School. (From Wallace, R. E., U.S. Geol. Surv. Circ., 701, 1974, 10.)

from Cajon Pass to the city of Riverside, and the other extends from Coyote Mountain to the community of Anza. Both are (1) approximately 40 km long, (2) void of fault creep, and (3) currently experiencing a sequence of small quakes. These researchers believe that strain has been accumulating in the gaps and that the next moderate shocks will occur there.

D. Pattern Recognition

Historical seismicity data reveal that $M_s \geq 6.5$ earthquakes in Central Asia (36° to 44°N, 60° to 80°E) occur in certain "disjunctive knots" or areas where major faults (active since the Neogene Period) intersect. The knots occupy only a small percentage of the total area.[661]

Gelfand et al.[661] developed a computer program involving a *pattern recognition algorithm* to automatically categorize all knots regarding their potential as future sites for strong earthquakes. Input data in binary form included certain geomorphological characters for each knot (e.g., type of fault junction, number of faults, length of major faults, distance from faults separating mountain countries) and epicenters of strong earthquakes (1885 to 1971). Knots were classified as:

1. Dangerous — where strong earthquakes have occurred
2. Potentially dangerous — where strong shocks are unknown but possible
3. Nondangerous — where strong earthquakes are not possible

The results were most promising. For example, from a historical perspective, the pattern recognition algorithm identified all knots where strong shocks had occurred between 1911 and 1971. Six knots were categorized as potentially dangerous for future (post 1971) strong earthquakes.

The group did not use data sets such as microearthquake histories, various geophysical anomalies, crust and upper mantle structures, and the tectonic history. Although they do not rule out the use of these data for other areas, for the present, at least, they believe they can predict earthquake sites in Central Asia from pre-existing geomorphological descriptors and histories of strong earthquakes.

Press and Briggs[662] applied standard geologic data in binary form to a pattern algorithm to identify earthquake-prone areas in California and Nevada. Experimental attempts at predicting earthquake sites showed positive results, and several predictions have been made for future sites.

II. EARTHQUAKE PRECURSORS

Seismological, geophysical, and geodetic methods are being used to isolate and monitor potential precursors or nonlinear changes in the physical state of the earth prior to the occurrence of earthquakes. This section describes the precursors that offer potential for predictions in a single or multiple seismic region(s).

A. Fault Creep

As was previously discussed, fault creep is currently found along certain segments of the San Andreas and branch faults. Sometimes within a creep zone, near-surface *patches* or gaps become stuck or locked and subsequently experience stick-slip events once accumulating strain exceeds the frictional resistance of the locked patches. Based upon these parameters, Wesson et al.[663] maintain that it may be possible to formulate a prediction framework for a 200-km section of the San Andreas fault between Cholame and Corralitos.

Using a *steady-state seismic slip model,* Bufe et al.[664,665] of the USGS predicted a small earthquake ($M_L = 3.2$) on a 9-km segment of the Calaveras fault approximately 15 km southeast of San Jose. Basically, the model is comprised of the following elements:

1. Strain is stored in the vicinity of a stuck patch that is tectonically driven at a constant rate within a "field" of constant fault creep.
2. The patch experiences stick-slip when the strain accumulates such that the stress across the patch exceeds the static frictional resistance.
3. The stick-slip interval is the time span required to reestablish the stored strain released in the previous quake; microearthquake activity can delay the interval.

Based upon these model parameters, in October 1976, a $3 \leq M_L \leq 4$ earthquake was forecast at 37°17' ± 2'N, 121°39' ± 2'W within a 48-day time window commencing on January 1, 1977. The earthquake occurred on December 8, 1976 — 24 days before the window was to commence. However, the epicenter (37°16.1'N, 121°38.1'W) and magnitude ($M_L = 3.2$) fell within the predicted ranges.[665]

Another shock of the same magnitude range has been predicted for this patch. If the slip is steady, the quake is forecast for early July 1977, but if there is above average, interim microearthquake activity within the patch, the shock is expected to occur sometime in August 1977. Time-window parameters will be refined as July approaches.[666]

Bufe et al.[665] suggest that because of the elongate shape of the patch on the Calavaras fault, their prediction model may be applicable to major strike-slip faults such as the San Andreas and North Anatolian.

Anomalous creep episodes have preceded several small- to moderate-sized earthquakes along the central section of the San Andreas fault. By using data from the USGS creepmeter network (Figure 21 in Volume I, Chapter 2), Nason and Tocher [667] discovered an increase in creep movement before two earthquakes ($M_L = 5.6$ and 5.5) in April 1961 near Hollister. The average rate of creep had been 1.2 cm/year prior to 1958, but by 1959 and 1960, the rate increased to 1.9 and 2.0 cm/year, respectively. The two shocks occurred on April 9, 1961. In addition, a 3-mm creep event preceded, by approximately 20 hr, the Melendy Ranch $M_L = 4.6$ earthquake (near San Juan Bautista) of September 4, 1972.[668]

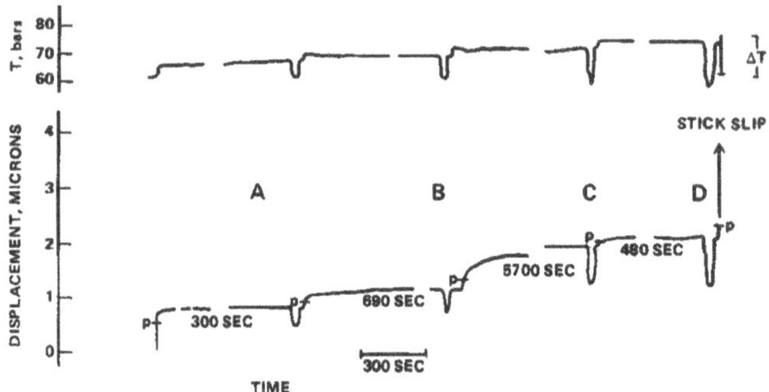

FIGURE 6. Creep before stick-slip on Westerly granite. The shear stress (top tracing) is increased or decreased in small steps, and the resulting displacement (bottom tracing) is recorded. Breaks in the tracing represent compressed time intervals. Displacement up to the tick marks (P) occurred while stress was being increased; displacement occurring after the tick mark is creep. (From Johnson, T. L., *J. Geophys. Res.*, 80, 2604, 1975. Copyrighted by American Geophysical Union. With permission.)

The laboratory studies of Scholz et al.[648] and Johnson[669] indicate that stick-slip motion was *always* preceded by a small amount of creep along pre-existing sliding surfaces involving Westerly granite, Twin Sisters dunite, and Spruce Pine dunite. The rock samples were tested in a biaxial frame where loads could be independently applied to the samples in the horizontal (F_H) and vertical (F_V) directions.

The creep observations indicated that the shear stress needed to initiate stick-slip was approximately equal to that required to cause creep; stable sliding may, therefore, be a fundamental part of the stick-slip process. Figure 6 depicts Johnson's[669] study of creep episodes prior to stick-slip on Westerly granite.

According to Scholz et al.,[648] stable sliding was directly scaled with the strain rate. If transposed to the field where the earthquake recurrence rate was 100 years, they believe creep could become "observable" 25 to 50 years before the shock while slip would increase to a high rate in the few years preceding the earthquake. Scholz et al. note that it is conceivable that the creep now observable in central California is this process in operation.

B. Foregoing Seismic Activity

Several earthquakes have been preceded by foreshocks, but if such activity is to be used as a prediction parameter, a means must be available to distinguish foreshocks from a region's "normal" seismic activity for which no primary earthquake follows. Suyehiro and Sekiya[670] believe that it may be possible to identify foreshocks by an abnormally low b-coefficient in the Gutenberg-Richter magnitude-frequency equation (Equation 32 in Volume I, Chapter 2).

According to Suyehiro and Sekiya, the value of b varies among seismic regions, but its value appears to be essentially unchanged for ordinary and aftershock events for the same region. For example, 25 foreshocks and 173 aftershocks were recorded in association with a January 1964 earthquake ($M_L = 3.3$) in Japan; b-values were 0.35 ± 0.01 and 0.76 ± 0.02, respectively, in the relation of frequency and magnitude (Figure 7). The 0.76 aftershock value agrees with that for the region's ordinary activity. Significantly lower b-values have been also associated with a 1967 earthquake ($M_L = 5.1$) in the same area of Japan,[670] the great Chilean earthquake ($M_S = 8.5$) of May 22, 1960,[671] and a $M_S = 3.0$ event that occurred near Fairbanks, Alaska, on November

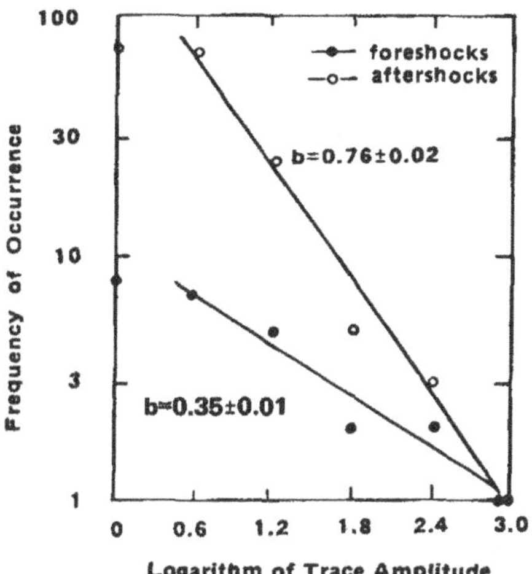

FIGURE 7. Relation between occurrence frequency and magnitude for foreshocks and aftershocks associated with a M_L = 3.3 January 1964 earthquake in central Japan. (From Suyehiro, S. and Sekiya, H., *Tectonophysics*, 14, 220, 1972. With permission.)

12, 1970.[671a] For the latter event, the b-value began to decrease 3 days before the shock, reaching a minimum level 1 day before the earthquake. Suyehiro and Sekiya[670] believe that if the low b-coefficient was found to be universal, it would be a simple technique to warn of ensuing main shocks.

Certain moderate, large, and great earthquakes, without foreshocks, have struck in regions characterized by a higher than normal level of seismic activity. For instance, according to Tocher,[672] the great 1906 San Francisco earthquake was predated by an increased level of moderate shocks in the bay region — about one per year for several decades.* Tobin and Sykes[673] report that, in the 10-year period preceding the great 1964 Alaskan earthquake, there was increased seismicity near the ends of the 1964 rupture zone, with one cluster of events coinciding with the approximate epicenter of the main shock.

Upon relocating epicenters of $M_s \geq 5.0$ events (1939 through 1967) along the North Anatolian fault in Turkey (Figure 4), Dewey[655] found that rupturing for large earthquakes commenced in regions of small and moderate earthquake activity and extended into areas that had a below normal level of activity — areas that were thought to be highly strained.

Kelleher and Savino[674] investigated the seismic preconditions for several large earthquakes associated with strike-slip and thrust faulting along the northwestern, northern, and eastern margins of the Pacific Basin. Two preconditions characterized most of the earthquakes. First, rupture-zone interiors remained essentially aseismic until the time of the main shock. Second, when there was prior seismicity, it was usually restricted to epicenter regions of the ensuing main shocks and/or to the edges of the pending rupture zone. There is some evidence to indicate that the level of foreshock activity appears to increase in the epicenter area as the time of the main shock approaches. Such patterns may be explainable by the dilatancy mechanism (discussed later in this chapter).

* This activity was almost entirely in the surrounding region and not along the 1906 rupture zone.

Their detailed analysis also indicates that a major earthquake may not be predated by anomalously high activity. For instance, premonitory activity for the July 30, 1972 Sitka, Alaska earthquake (M_s = 7.6) was essentially nonexistent from the mid-1960s. A similar pattern has been associated with several earthquakes in the People's Republic of China.[675]

Ohtake et al.[676] investigated the predate seismic histories associated with 11 large (M_s = 6.0 to 7.5), shallow earthquakes in Mexico and South America; in most cases, marked changes were recognized in the temporal seismicity pattern. Activity markedly decreased about 1 to 2 years in the region of the pending main shock, followed by a resumption of activity just prior to the main earthquake. Based upon this pattern, they believe there is a possiblity of a future earthquake near Oaxaca, Mexico.

Wesson and Ellsworth[677] analyzed foregoing seismicity for several moderate California earthquakes representing a variety of seismic (fault) environments:

1. Kern County; M_L = 7.7; June 21, 1952
2. Watsonville; M_L = 5.4; September 14, 1963
3. Corralitos; M_L = 5.0; November 16, 1964
4. Parkfield-Cholame; M_L = 5.1, 5.5; June 28, 1966
5. Corralitos; M_L = 5.3; December 18, 1967
6. Borrego Mountain; M_L = 6.4; April 9, 1968
7. Santa Rosa; M_L = 5.6, 5.7; October 2, 1969
8. San Fernando; M_L = 6.4; February 9, 1971
9. Bear Valley; M_L = 5.0; February 9, 1972

In every case, the main shocks occurred in areas characterized by a relatively high number of small magnitude (M_L <5.0) events. Additionally, in most of the examples, foregoing activity was concentrated near the epicenters of the impending main shocks rather than along adjacent parts of the same fault zone or other nearby faults that were equally suspect for moderate earthquakes based upon geologic evidence.

To explain these areal patterns, Wesson and Ellsworth proposed that stress is unevenly distributed along the fault zones. As the general stress level increases (most likely by plate motions), small segments reach the failure threshold (local stress concentrations) before a larger segment fails. They suggest that it may be possible to predict moderate and large earthquakes in California by monitoring small-magnitude earthquake activity along fault zones with expanded seismograph networks.

Sadovsky et al.[678] report on the use of forerunner seismic activity for long-range predictions in certain regions of Middle Asia. For example, if small shock activity decreases in a 5- to 10-year period or is stable for a particular region, a strong earthquake is thought to be unlikely for at least 5 more years. However, if the seismicity level increases within a 5- to 10-year interval, the region is classified as active; the possiblity would then exist for strong earthquakes during the next 10-year period.

During the Matsushiro, Japan earthquake swarm (1965 to 1967), numerous microearthquakes were recorded prior to larger events, and without exception, they were clustered in the epicenter regions of the larger earthquakes. This forerunner pattern was successfully used for long-range warnings (comparable to long-range weather forecats) issued by the Japanese government.[679] Sykes[680] summarizes this effort:

> One of the outstanding features of the Matsushiro swarm was the gradual enlargement with time of the fault region experiencing earthquakes. It was found that microearthquakes tended to migrate into a new region along the fault zone prior to the occurrence of moderate-size earthquakes a few months later. Based on such observations, warnings were issued that moderate-size earthquakes could be expected within a few months. One of several earthquakes predicted successfully in this way was filmed by cameramen who set up their equipment in advance.

C. Vertical Crustal Deformation

Anomalous surface elevation changes in epicentral zones have predated a number of earthquakes in Japan, the U.S., and the Soviet Union. The earliest accounts are from coastal areas along the Sea of Japan. For example, about 1 m of uplift reportedly preceded the 1793 Agrgasawa and 1802 Sada earthquakes by 4 hr, and the ground rose from 1.8 to 2.0 m some 9 hr before the 1872 Hamada earthquake. Regarding the latter event, weather observers noted that prior to the shock a large rock, normally separated from the island by a 2-m deep channel at low tide, could be reached by a dry route.[681]

Vertical crustal deformation is now usually measured by *tiltmeters* and *repeated leveling (releveling) surveys* or by *tide gauges* in coastal areas. Tiltmeters and tide gauges can provide a continuous monitoring capability, whereas releveling produces noncontinuous data sets because the surveys are normally conducted at one year to several year intervals over pre-established bench-mark networks.

A common type of tiltmeter currently used in the U.S. is a biaxial, electronic-bubble model that was developed for Minuteman III missle guidance systems.[682] Its application to prediction studies has been described by Bacon et al:[683]

> The bubble sensor . . . is similar in principal to the ordinary carpenter's level. The sensor has 4 evenly spaced platinum electrodes protruding into the fluid chamber. By measuring an electrical signal applied to each of 2 pairs of opposed electrodes, a reading can be obtained on 2 axes — 1 in a north-south and the other in an east-west direction. The data from the 2 axes, which are recorded on a chart recorder, can be combined to give the true direction and amount of tilt at any given time.

These instruments have a 10^{-7} to 10^{-8} radian sensitivity and are installed in shallow boreholes to minimize the effects of temperature, moisture, and atmospheric pressure variations (Figure 8).

The USGS has operated tiltmeter networks (1) along an 85-km segment of the central San Andreas fault since mid-1973 (Figure 9), (2) around the Los Angeles Basin and along the San Jacinto fault since 1975, and (3) along the Palmdale segment of the San Andreas fault, in the southern part of the state, since 1976. In most cases, tiltmeters are spaced approximately 5 to 6 km apart on alternate sides of a fault and from 1 to 4 km away from the fault. The California Division of Mines and Geology operates several tiltmeters in various parts of the state.

One of the first examples of recognized predate crustal deformation, based upon multiple leveling surveys, was associated with the June 16, 1964 Niigata, Japan earthquake (M_s = 7.5). In 1965, Tsubokawa et al.[684] analyzed five pre- and two post-earthquake leveling surveys. Crustal movements from 1898 to the early 1950s were essentially linear, with extremely small movement rates. However, the rate of movement started to increase in the epicentral region about 1955 and continued until the time of the earthquake; uplift in 1958 totaled 5 cm. The ground underwent rapid subsidence during and immediately after the earthquake.

Repeated leveling surveys have been completed by Boulanger et al.[685] in the Garm and Alma Ata regions of the Soviet Union. Even though no more than two levelings were made in any one year, there was a relation between the premonitory displacement of bench marks and several strong earthquakes. However, they point out that the frequency of measurements must be increased to perhaps once a month to fully ascertain the importance of preseismic uplift in these two active regions.

Vertical crustal movements preceded the February 9, 1971 San Fernando earthquake (M_s = 6.5).[686,687] After the event, Castle et al.[686] analyzed elevation data for bench marks located along the Central Transverse Ranges of southern California and found that the mountain system had undergone significant elevation changes in 1961, 1964, 1965, 1968, and 1969. Uplift totaling 20.7 cm occurred approximately 28 km northeast

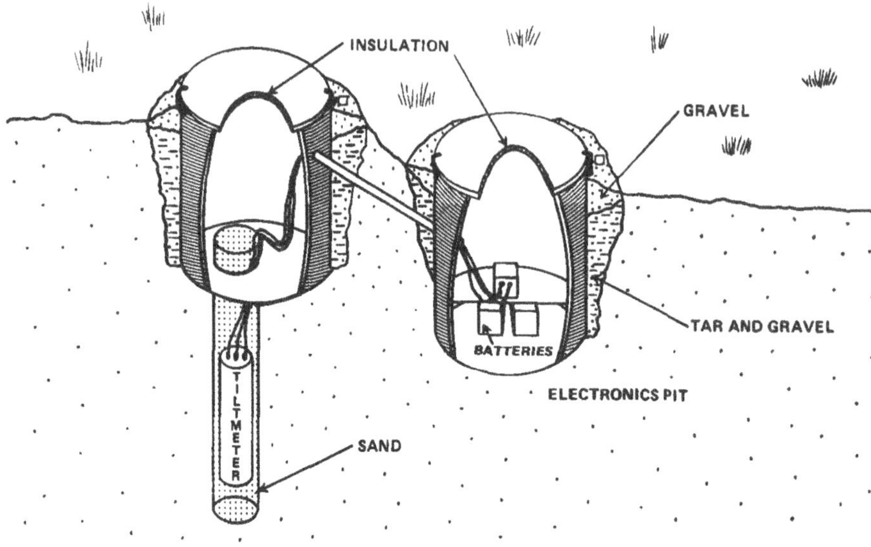

FIGURE 8. Typical U.S. Geological Survey *tiltmeter* installation. Pit at left is 1 m in diameter, 2 m deep,and lined with a fiberglass culvert surrounded by a layer of tar and gravel to prevent seepage. The tilt sensor is packed in a fine-grained silica sand near the bottom of a 15-cm diameter, 2 m-deep borehole lined with steel tubing and closed at the lower end. The recording and transmission electronics are located to the right in a similar pit. (From U.S. Geological Survey, *Earthquake Inf. Bull.*, 6, 17, 1975.)

of the epicenter between 1961 and 1964; between 1968 and 1969, the ground was elevated 7.8 cm 10 km west of the epicenter.[687] According to Castle et al.,[686,687] the overall movements may be interpretable as the product of continuing north-south contractional strain, while the more episodic movements may result from deep-seated creep events along the San Fernando fault and/or dilatancy.[686,687]

The first successful application of a tiltmeter for detecting predate crustal deformation occurred during the Matsushiro earthquake swarm in Japan. Hagiwara and Rikitake[688] conducted a water-tube tiltmeter observation program in an underground vault at the Matsushiro Seismological Observatory. Significant, long-term inclination changes were noted during swarm activity, and short-term tilting was detected a few hours before magnitude 5 earthquakes. The government issued successful warnings to the public in April and August 1966 that were based, in part, upon these tilt anomalies.

In California, changes in the direction of surface tilt have preceded several earthquakes or groups of earthquakes. Similar to the findings of Hagiwara and Rikitake,[688] Wood and Allen[689] discovered that two tiltmeters recorded long- and short-term anomalous tilt activity premonitory to three earthquakes (M_L >4.0) on June 12, 1970. The earthquakes were associated with the Pleasanton fault near Danville, and the instruments were located in vaults at the University of California, Berkeley, and the Presidio, San Francisco — 25 and 45 km west of the epicenters, respectively.

Data analysis indicated a constant tilt of 8×10^{-10} radian/hr for 1 month and an accelerated tilt change during the day preceding the three shocks. The 1-day tilt change was most pronounced at the station closest to the epicentral area. Wood and Allen attributed the 1-month anomaly to the build-up of strain originating at a mantle-crust depth and the 1-day anomaly to the initiation of crustal failure near the surface.

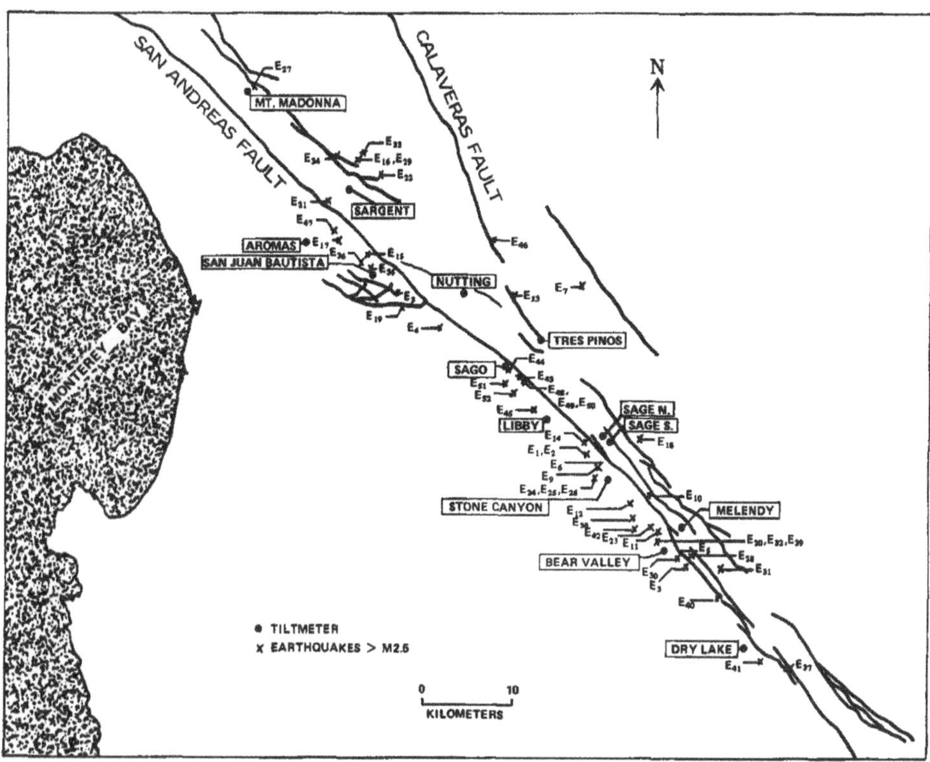

FIGURE 9. Tiltmeter sites (1974) along an 85-km section of the central San Andreas fault system and earthquakes over M_L = 2.5 that have occurred in the vicinity of the network between July 1973 and March 1974. (From U.S. Geological Survey, *Earthquake Inf. Bull*, 6, 16, 1975.)

Using data from the USGS tiltmeter network, Johnston and Mortensen[690] found premonitory short-term tilt direction changes for more than ten earthquakes or groups of earthquakes (M_L = 2.5 to 4.4) along the central San Andreas fault during a 7-month period in 1973 and 1974. *Each* time this anomalous activity occurred, seismic activity *always* followed. Tilt and earthquake data for the Nutting tiltmeter site are presented in Figure 10. Of special interest is the tilt record associated with the January 10, 1974 earthquake. Note that from December 26 to the time of the shock, the direction of tilt reversed from southeast to northwest, or towards the direction of the impending earthquake. The rotation continued after the event, terminating once the rotation reached 360°.

The following passage summarizes the multi-year findings of Johnston and Mortensen.[691]

> The tilt effects have been recorded simultaneously on up to four independent stations. No indication of a sudden increase in the rate of tilting immediately before an earthquake has been observed.
>
> The general character of the tiltmeter data indicates that long-term systematic tilting up to 2×10^{-6} radians per month occurs in the vicinity of the array. This long-term tilting has been consistent at some sites while at others the trend has changed direction markedly. In general, the secular tilting increases with seismic activity in the surrounding region probably as a result of long-term tectonic effects on the fault.
>
> Short-term tilt changes with durations of up to 1 month are superimposed on the long-term systematic tilting. These short-term changes are associated with numerous 2.5 to 4.4 earthquakes that have occurred in the vicinity of the tiltmeter array . . . . The short-term tilt changes are apparently limited to within a radius of 10 source dimensions from the earthquake's epicenter . . . . It is the change in tilt direction that appears to be most directly related to the subsequent earthquake. Changes in the amplitude of tilting also occur, *but are not as clearly defined.*

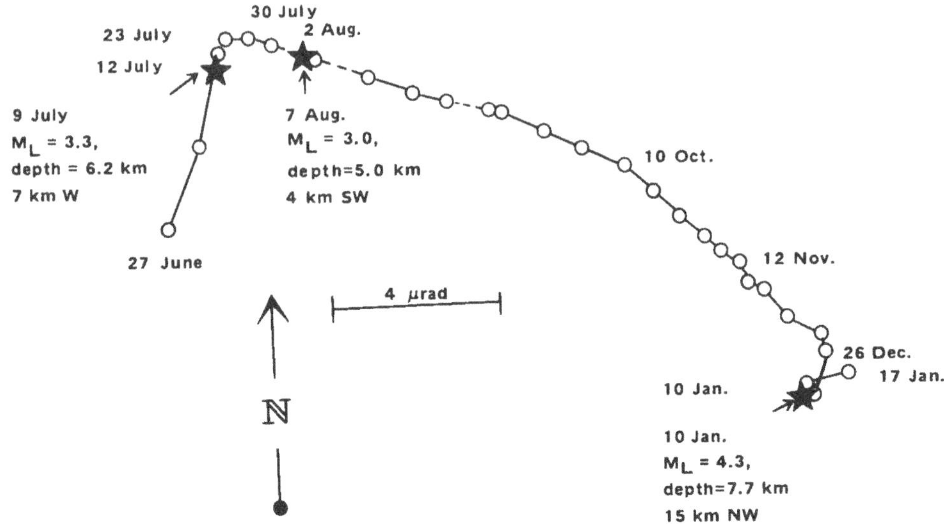

FIGURE 10. Cumulative weekly mean tilt vectors (circles) from 27 June 1973 to 17 January 1974 for the Nutting site, 7 km southwest of Hollister. Tilts toward the north and east are along the positive ordinate and abscissa, respectively. (From Johnston, M. J. S. and Mortensen, C. E., *Science*, 186, 1031, 1974. Copyright 1973 by the American Association for the Advancement of Science. With permission.)

In early 1975, Stuart and Johnston[692] discovered premonitory surface tilting in the focal regions of three earthquakes (M_L = 3.2, 3.3, 4.2) along the same segment of the San Andreas fault. The nature of the tilting agreed reasonably well with the theoretical tilt due to a small amount of right-lateral dislocation. However, a few of the observations exhibited radial symmetry about an epicenter, which indicates domal uplift.

On November 28, 1974, a M_S = 5.2 earthquake struck near Hollister on an extension of the Busch fault, which is situated between the Sargent and Calaveras faults. The earthquake was not totally unexpected because significant tilt and magnetic field anomalies had been detected before the shock. This marked the first time in the U.S. that two potential precursors were observed premonitory to the actual occurrence of an earthquake. The magnetic anomaly is described in a later section.

According to Mortensen and Johnston,[693] tilt perturbations caused by pre-earthquake distortion commenced 36 days before the earthquake at four stations, all within 15 km of the epicenter. The tilt vectors changed with time, reaching values greater than 10^{-6} radians. Because two strongly suggestive precursor-type phenomena had been discovered, several USGS scientists involved in prediction research speculated informally on the distinct possibility of an imminent shock at a meeting on November 27 — the earthquake struck the next day.

Johnston and Stuart[694,695] recently used surface-tilt data to successfully predict a small earthquake in central California. On December 30, 1976, they stated that a M_L = 3.5 shock would occur on the San Andreas fault within 15 km of the Howard Harris Ranch (near Hollister) between January 5 and 15 (bracketed to all of January). A M_L = 3.2 earthquake struck on January 6 approximately 10 km south of the ranch on the San Andreas fault.

Based upon tilt observations along the central San Andreas fault since mid-1973, the prediction model equates (1) magnitude with how many tiltmeters in the USGS array record the vector changes and (2) time of occurrence with the anticipated magnitude: > number of tiltmeters > magnitude and > magnitude > advance warning. The number of tiltmeters recording the anomalous activity determines the scale of the deformed region along the fault, which then makes it possible to estimate magnitude. Three successive tiltmeters, separated by distances of 4 to 5 km, recorded anomalous activity for the January 6 earthquake. Based upon a horizontal slip model, the directional patterns of tilt were used to locate the potential epicentral region. The prediction was not announced because of the earthquake's potential small size. However, had the evidence indicated a potentially dangerous shock, a public broadcast would have been made to local residents by the USGS.[694]

Preliminary results from the USGS Los Angeles Basin-San Jacinto fault tiltmeter array indicate that the Table Mountain instrument, located adjacent to the San Jacinto fault, recorded anomalous tilting premonitory to a $M_L = 4.9$ earthquake.[696] Accelerated tilting to the northwest commenced about May 25, 1975 and continued until July 16, at which time the instrument went off-scale. On August 1, the shock occurred a few kilometers south of the tiltmeter site.

Data from Rockwell International's biaxial tiltmeter, located in Anaheim, California, indicate that anomalous tilts have preceded some, but not all $M_L \geq 4.0$ earthquakes occurring within a 500-km radius of the detector site.[697,698] Buckley et al.[697] maintain that the observation site has an favorable location to detect precursor tilts to the south and east or in the crustal block bounded by the San Jacinto and Newport-Inglewood faults and latitudes 32° and 34°N. Preliminary findings indicate that a southeast down tilt, exceeding 1 sec arc and persisting for two days, is premonitory to $M_L \geq 3.0$ earthquakes occurring within this block.

According to Sykes and Raleigh,[675] the monitoring of tilt comprises a large part of the prediction effort in the People's Republic of China, where tilt changes have predated several earthquakes. For example, the Yunnan Province earthquake ($M_L = 4.8$) of July 17, 1972 was preceded by anomalous tilting that persisted for 15 days. This earthquake was predicted publicly.[675]

The previous tilt studies were concerned with primary shocks, but Sylvester and Pollard[699] believe that tilting might be a precursor for potentially dangerous aftershocks. Approximately 5 hr before two of the stronger aftershocks of the February 9, 1971 San Fernando earthquake, there was drift in a theodolite plate-level bubble toward the Sylmar segment of the San Fernando fault (Figure 5 in Volume I, Chapter 3). A total of six quadrilateral surveys had been completed to monitor postearthquake strain along the fault. On one of the surveys, the theodolite tilt was observed before the two aftershocks, but neither tilting nor aftershocks were associated with the other five surveys.

A large uplifted land area has been discovered recently astride the San Andreas fault near Palmdale and has become known as the "Palmdale bulge"[700-703] (Figure 11). The uplifted area is astride the portion of the fault that was last struck by a large earthquake in 1857. Its significance remains a mystery, but because similar crustal expressions have predated several earthquakes in different seismic environments, the bulge is being monitored closely by the USGS and the California Division of Mines and Geology. Various instrument arrays have been deployed to search for potential precursors (e.g., seismic, electrical resistivity, and magnetic field anomalies). Based upon experiences in the People's Republic of China, certain animals (e.g., pocket mice and kangaroo rats) are being monitored for unusual activity that may predate an earthquake. The USGS received a special $2 million appropriation for Fiscal Year 1976 to 1977 to conduct several types of investigations in the area.

Characteristics of the bulge include the following:[700]

1. The uplift was discovered in 1976 by a releveling survey; it is believed that the swelling commenced about 1960 near the junction of the San Andreas and Gar-

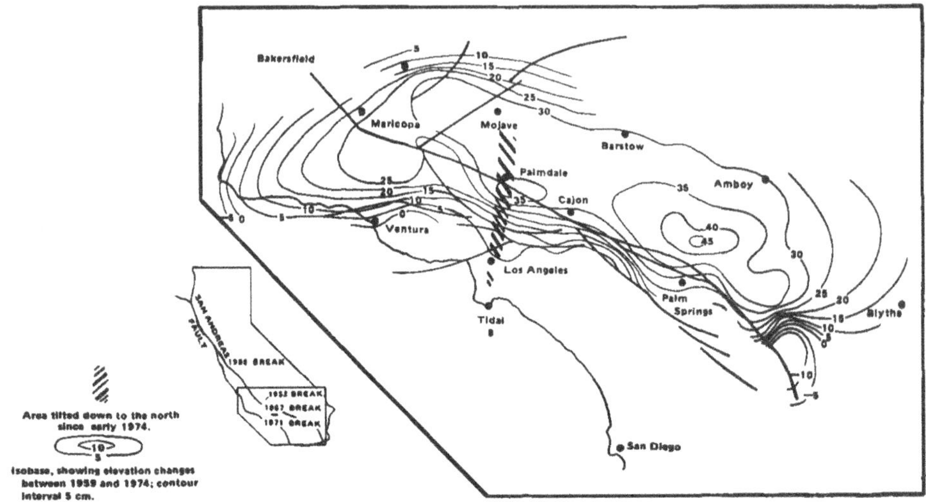

FIGURE 11. *Palmdale bulge* astride the San Andreas fault. (U.S. Department of the Interior News Release.)

lock faults. Between 1960 and 1976, the bulge was extended east-southeast and was thought to encompass about 12,000 km² at the time of discovery. Further analysis now indicates that the uplifted area totals 90,000 km² — extending 600 km from Point Arguello eastward to the Arizona border and 150 km from Los Angeles to Bakersfield (Figure 11).

2. At least part of the area has collapsed. The area centering on Palmdale rose approximately 35 cm between 1960 and 1974, but the area has subsided by about 18 cm since 1974 (Figure 11). This apparent cycle of uplift and subsidence may have occurred in the same area earlier in this century. Between 1897 and 1914, 30 cm of uplift occurred, but a 1926 releveling survey indicated collapse by essentially the same amount.
3. No earthquake or group of earthquakes can be clearly associated with this earlier episode of uplift and collapse. Therefore, the present pattern may or may not be a precursor of seismic activity.

D. Electrical Resistivity

The laboratory research of Brace and Orange[704-706] suggested the possibility of using anomalous electrical resistivity data for predicting earthquakes. They found that as the confining pressure or compressive stress was increased in water-saturated granite and other crystalline rocks, a specimen's resistivity to an induced electric current increased slightly (i.e., conductivity decreased) until about half the stress needed to cause brittle fracture was reached and then decreased until rupturing occurred. Just before failure, the drop in resistivity was approximately an order of magnitude below the normal resistivity level.

Both active and passive *variometers* are being used to monitor potential changes in the earth's electrical resistivity. An active variometer utilizes a weak source of direct or alternating current, and two common configurations are used: *dipole-dipole* and *Wenner.* The former array is emplaced just below the surface (e.g., 2 m) and consists of two pairs of electrodes: a transmitting or source dipole and a receiving dipole. The horizontal spacing between the source and receive dipoles is usually several kilometers (e.g., 5 to 7 km), enabling a crustal depth of about the same dimension to be monitored. A direct current is passed into the ground by the source dipole. At the receiving dipole, a record is made of the voltage diminution which is proportional to the input current, the ground's resistivity between the dipoles, and to some coefficient determined by the geometry of the array. Therefore, for a given geometry, the ratio of the voltage change to the input current is proportional to resistivity.[707,708] The dipole-dipole array is used commonly in the U.S. and the Soviet Union.

The Wenner configuration is comprised of four electrodes buried in a straight line, with a relatively narrow spacing between electrodes (e.g., 200 m). An alternating current is transmitted by the two outer electrodes, and the apparent resistivity of the ground is proportional to the voltage recorded by the two interior electrodes.[708] The Wenner array has been adopted for prediction research in Japan.

The passive sounding technique monitors the earth's *telluric current.* This long-wave form of energy is thought to be produced by the solar wind impinging upon the magnetopause. The energy reaches the earth's surface in the form of micropulsations,[709] and moves through the crust (with deep penetration) in continental-sized "sheets."[683] Precursor-type monitoring programs are operative along the eastern shore of Kamchatka in the Soviet Union,[680,710] several locales in the People's Republic of China,[675] eastern Canada,[711] and California.[683,712]

The following is a description of the California Division of Mines and Geology's five-station telluric measuring network along the San Andreas fault between the Carrizo Plain and Parkfield.[683]

> A typical station consists of 4 nonpolarizing electrodes, each about 1/4 mile . . . (402 m) apart, with interconnecting copper wires. Each electrode contains a probe in a porous container filled with a copper sulfate solution and buried in the ground. The copper sulfate solution improves the electrical contact between the ground and the copper probe and helps to minimize local spurious potentials. Potential differences between the probes are recorded on a chart recorder.
>
> Simultaneous measurements of telluric current magnitude changes or 'events' made at 2 or more stations were begun in 1974. Successive measurements have proven that a constant ratio of the recorded magnitude of these telluric current 'events' exists among the several stations . . . A local change in the Earth's resistivity, which may precede an earthquake near one or more of the stations, may be expected to upset this constant ratio and thus provide an indication of a forthcoming earthquake.

Although noise problems caused primarily by precipitation, ocean tides, and magnetic storms have reduced the statistical validity of the data, short-term variations in the telluric field may have predated several earthquakes in the People's Republic of China[675] and the Soviet Union.[680,710]

Resistivity changes involving an artificially introduced electric current were first detected in crustal rocks comprising the seismically active Garm region in the Soviet Union.[713] During a monitoring program from January 1967 until March 1971, large thrust-type earthquakes were preceded (1.5 to 2.0 months) by a diminution of resistivities ranging from 15 to 18% below normal values (Figure 12).

Barsukov[713] believed that the diminished resistivities might be explainable by an increase in the volume of pure fluid in hypocenter zones. He reasoned that in such a zone, fluid was present in the pores and cracks of the mountain rocks. With pore-pressure enhancement, there was an accompanying reduction in the shear strength of the rocks which would lead eventually to displacement and the earthquake. Coincidental with the diminishment of rock strength was a reduction in electrical resistivity because of an increase in the volume of the electroconducting liquid. Therefore, premonitory to an earthquake, the increased volume of liquid in the hypocenter zone would produce a diminished resistivity to an electric current.

In 1973, a dipole-dipole array was established across a segment of the central San Andreas fault. During its first year of operation, Mazzella and Morrison[714] discovered a 24% change in resistivity premonitory to a M_L = 3.9 earthquake (June 22, 1973)

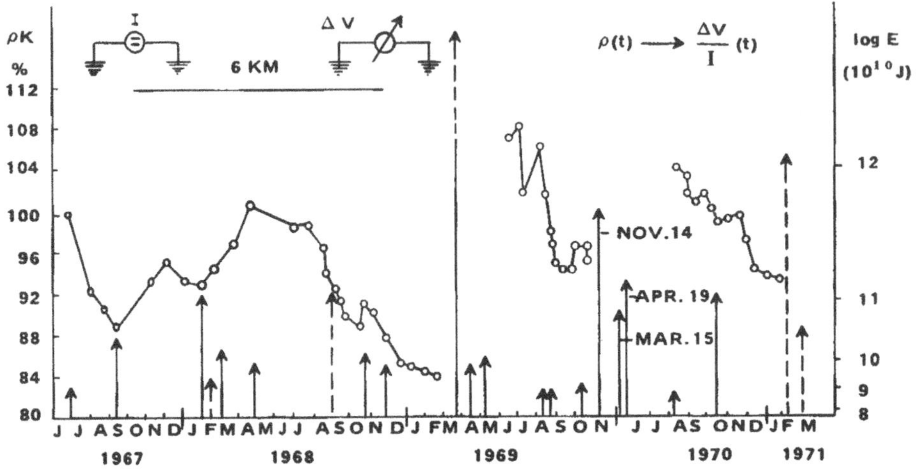

FIGURE 12. Variation of electrical resistivity and the occurrence of large thrust-type earthquakes in the Garm region of the Soviet Union. The ordinate represents values of resistivity as a percent of the initial value. The vertical arrows show the moments when earthquakes occurred while the right-hand scale shows the logarithm of their energy. (From Barsukov, O. M., *Tectonophysics*, 14, 274, 1972. With permission.)

near Hollister. The resistivity sequence was similar to that reported by Brace and Orange:[704-706] initial decrease followed by an increase. The precursor activity persisted for 60 days. Although the data are less definite, resistivity variations along the same fault segment may have also predated an October 8 to 13, 1973 swarm sequence and a $M_L = 2.6$ shock on December 14, 1973. Because of the success of this program, the University of California, Berkeley, array has been expanded to cover 45 km of the fault south of Hollister.[715]

A Wenner array variometer, located approximately 60 km south of Tokyo, has recorded step-like resistivity changes (strain steps) for a number of distant earthquakes (sometimes exceeding 1000 km) centered off the coast of Japan. In a few instances, precursory changes in resistivity were recorded 1 to 7 hr before the earthquake struck. The precursory signals were enhanced by eliminating long-period changes, due mostly to tidal loading, by a numerical filtering technique.[716-718]

E. Tectonomagnetic Effects

The thesis that a correlation exists between seismic activity and anomalies in the earth's magnetic or *geomagnetic field* (i.e., *tectonomagnetic effects*) dates back to the late 1800s in Japan. For example, Rikitake[182] found a report in classical literature where pieces of iron attached to a horseshoe magnet in an optician's office in Yedo (now Tokyo) fell to the floor about 2 hr before a destructive earthquake in 1885. In other instances,[182,719] geomagnetic anomalies, as recorded by *magnetometers*, were thought to have been forerunners to several earthquakes. However, Rikitake[719] believes that the intensity data contained large errors because there was no correction for nonlocal geomagnetic changes (e.g., magnetic storms, diurnal variations).

One of the most recent examples of a magnetic disturbance preceding a seismic event was that associated with the March 27, 1964 Alaska earthquake ($M_S = 8.5$). A magnetometer operating in the community of Kodiak was serving as a ground-based monitor for an airborne magnetic survey. Following the earthquake, Moore[720] examined the *magnetogram* and discovered that the intensity of the geomagnetic field had increased by 100 gammas 66 min before the earthquake. Moore speculated that the anomaly may have resulted from the rocks undergoing a stress change before the shock. Like earlier accounts, the discovery was by chance and not part of a program designed to search for tectonomagnetic effects.

Programs to test the possible link between earthquakes and geomagnetic anomalies had their beginning in 1922 when Wilson,[721] using laboratory data, proposed that stress changes in the crust should cause variations in the local geomagnetic field, and hence, these variations might be useful for monitoring earthquake activity. Beginning in the 1950s, this view was fortified by a number of laboratory experiments[722-725] and theoretical studies.[726-733] Laboratory data revealed that when rocks containing magnetite were subjected to uniaxial compression, the specimen's *magnetic susceptibility* and *remanent magnetism* decreased in the direction of compression and increased to a lesser extent at right angles to the applied stress. This stress-dependent behavior is called *piezomagnetism* or *seismomagnetism*.

Because of favorable laboratory and theoretical results, magnetometer networks were established in several earthquake-prone areas to determine if subsurface tectonic stress changes could be identified by measurable changes in the intensity of the local geomagnetic field. Two types of total field magnetometers have been used: (1) the *optically pumped rubidium vapor magnetometer* and (2) the more recently introduced *proton precession magnetometer*. The former records the field continuously and can detect intensity changes as small as 0.02 gamma. Proton precession magnetometers sample the field at short-interval cycles (e.g., every 30 sec) with a 0.25-gamma sensitivity. However, unlike the rubidium vapor magnetometers, they are practically drift-free, thus making it possible to search for long-term changes in the geomagnetic field.

In August 1965, Stanford University established the first network along a 150-km segment of the central San Andreas fault to search for possible piezomagnetic effects. The array consisted of five rubidium vapor magnetometers spaced at 30-km intervals. To identify any change in the local field (estimated to be between 0.001 and 10 gammas), micropulsations, which were ten to hundreds of times larger than the expected piezomagnetic effects, had to be removed from the data bases. Because micropulsations originate primarily in the ionosphere, it was assummed that adjacent sensors would record essentially identical effects and that the difference in their variations would be constant. Therefore, if a local magnetic disturbance occurred closer to one of the magnetometers, there should be a change in the otherwise constant difference.[734,735]

Breiner and Kovach[734,735] discovered six apparent changes in the geomagnetic field during the network's first several years of operation, and creep activity followed *each* change. In five cases, creep displacement (0.5 to 4 mm) occurred on the San Andreas fault near Hollister several tens of hours after the magnetic disturbances. In the sixth example, after the array had been reorganized into a higher density net, a magnetic event was recorded simultaneously by four adjacent magnetometers. A 4-mm creep episode followed in about 16 hr, and the area was struck by several small earthquakes 4 days later.

More recently, encouraging results have been obtained from the USGS's total-field, proton precision magnetometer networks.[736-740] One array established in 1973 is comprised of approximately 120 stations spaced at 10- to 15-km intervals along active faults in California and western Nevada. A resurvey technique is employed whereby local field data are collected at each magnetic station at 6-month intervals or more often if the seismicity increases. The observed values are reduced to those of a standard or reference station. The objective of this program is to search for long-lived variations in the geomagnetic field that might predate a large earthquake.[738]

A second array began continuous operation along an 80-km segment of the central

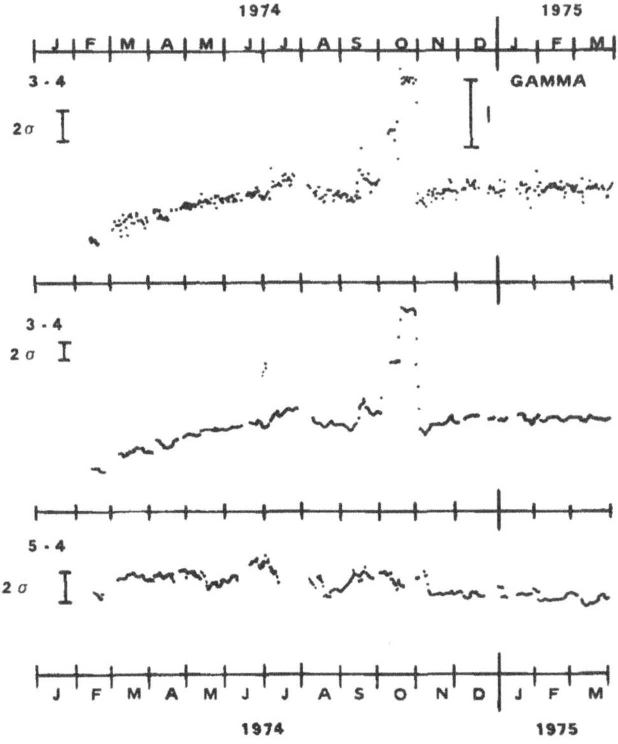

FIGURE 13. Plots of 1-day means (top record) and 5-day running means (bottom two records) of the total magnetic field differences for station pairs 3 and 4 and 5 and 4 from the U.S. Geological Survey's seven-station proton precission magnetometer network in central California. The error bars represent 2 SD calculated by using the 1- or 5-day mean values from March through July 1974 after removing the long-term trend. This period of time includes the largest ionospheric disturbances observed during 1974. The M_L = 5.2 Hollister earthquake occurred on November 28, 1974. (From Smith, B. E. and Johnston, M. J. S., *J. Geophys. Res.*, 81, 3558, 1976. Copyrighted by American Geophysical Union. With permission.)

San Andreas fault (the most active zone) in January 1974. The seven instruments operate synchronously, and the data are telemetered in digital form to Menlo Park for analysis in near real time.[738,741]

To date, the most definitive magnetic disturbance was associated with the M_L = 5.2 Hollister earthquake of November 28, 1974.*[,740] Significant changes in the local geomagnetic field were recorded at a site 11 km from the epicenter. The following intensity patterns for this station (i.e., station #3) were determined by Smith and Johnston.[740] The local field increased by 0.9 gamma from mid-February to late July, diminished slightly from August to mid-September, increased by 1.5 gammas during October and persisted at this level for 2 weeks, and then decreased by 1.8 gammas around November 1 (Figure 13).

Smith and Johnston[740] believe that the data can be explained by the piezomagnetic effect wherein the field changes represented stress changes in rocks close to the anom-

* It will be remembered from a previous discussion that premonitory tilting[693] was also observed before the 1974 Hollister earthquake.

alous station. Because of this successful observation, a larger magnetometer array is being considered for the San Andreas fault.

Proton precession magnetometer networks are also a part of governmental prediction research programs in the Soviet Union, Japan, and the People's Republic of China. Observation programs are being conducted in the Central Asian Republics of the Soviet Union.[742] Abdullabekov et al.[743] report that anomalous activity was associated with a February 1971 earthquake in the Tashkent region. The Japanese array consists of 12 stations, but the program is being hampered by extraneous natural and cultural noise at a level far exceeding the accuracy of the measurements. Standard deviations of from 2 to 5 gammas between adjacent stations are apprently being caused by (1) electric currents excited in the extremely heterogenous crustal rocks (nonuniform conductivity) by variations in the earth's external magnetic field and (2) stray electric currents being induced into the ground from the country's extensive electrical railway system.[182,744,745]

Although local anomalies were observed in the geomagnetic field during the most violent stage of the 1965 to 1967 Matsushiro swarm,[719,746] Rikitake[744,745] believes that because of the noise problems the geomagnetic method does not appear promising for predictions, with the possible exception of magnitude 7 and larger earthquakes. With this class of events, the seismomagnetic effect might be above the noise level.[744,745]

In the People's Republic of China, the vertical component of the geomagnetic field is monitored at 17 principal seismological observatories and a closely spaced net near Peking. Although little is known about the program, Sykes and Raleigh[675] were informed on a recent visit that anomalies in the differences between the Hung-shan and Peking field intensities preceded two earthquakes (magnitudes 4.4 and 5.2). In addition, they were told that a large increase in the differential field was observed 2 days before a 4.9 earthquake struck near Hsingai on June 6, 1974. With this and other precursor observations, personnel at the Red Mountain Observatory issued a successful public prediction.

F. Radon Emanation

Beginning with the April 26, 1966 Tashkent earthquake ($M_L = 5.2$) in the Soviet Union, anomalous changes in the amount of *radon emanation* form either water or soil has shown promise as a possible earthquake precursor. Radon is a radioactive gas that forms with alpha radiation as an initial product in the natural decay of radium. It has a half-life of 3.8 days.

In the early 1960s, water from a well in the Tashkent artesian basin was being analyzed on a regular basis to determine any possible changes in its chemical properties. The well was sited in what was to become the hypocentral region for the 1966 earthquake.[680] Ulomov and Mavaschev[747] found that the radon content of the water displayed a most interesting pre-earthquake history. Radon emanation increased for 5 years before the quake and then suddenly, 1 year before the shock, the radon exhalation leveled off. Increases in radon emanation were also detected before several of the aftershocks. More recently, radon anomalies in well water in the Tashkent region occurred several weeks before two earthquakes.[748]

Following a visit to the People's Republic of China in 1974, Sykes and Raleigh[675] reported that "some of the work in China on radon emanation as a precursory effect is to our knowledge as good as that being done anywhere in the world." Water samples withdrawn from several tens of wells and springs located near active fault zones are being analyzed by *electrometers* once or twice a day. An electrometer measures the flux of alpha particles and, hence, the radon flux.

Both increases and decreases in radon emanation have been observed before earthquakes in the People's Republic of China. For example, Sykes and Raleigh[675] were told that seismologists at the Red Mountain Observatory discovered anomalous increases in the radon content of water for five earthquakes (magnitudes 4.3 to 5.0) and a decrease for one 7.9 earthquake centered in the Pohai Gulf. For the smaller shocks, the anomalies were recorded at distances less than 110 km from the epicenters. The span of time between anomaly observation and earthquake occurrence ranged from a few days to several tens of days. For the 7.9 event, a strong decline in radon emanation was observed in several wells located 250 to 300 km from the epicenter.

Several radon measuring projects are operative in California. USGS[749-751] and University of California, Los Angeles (UCLA),[752] projects monitor radon emanation from soil air by the *Track Etch method*. This method makes use of dielectric film that is sensitive to alpha radiation. At each site, a piece of film is exposed to soil air in a shallow, capped hole (0.7 m deep) for approximately 1 week. The alpha particle track density on the exposed film is subsequently used as a measure for determining the amount of radon emanation. The USGS currently maintains 60 monitoring sites along several active fault traces in central California; 14 sites are monitored by UCLA personnel along a 25-km segment of the San Jacinto fault zone.

Large temporal variations in radon emanation have preceded by periods of weeks several M_L = 4.0 to 4.6 earthquakes in the USGS monitored region.[751] The anomalous behavior was recorded at sites nearby to each epicenter, and in every case, the radon emanation followed a common pattern: a rapid increase to a peak or plateau well above the background level with the quakes occurring shortly after the peak or plateau was reached. For example, 20 stations in the USGS array recorded a large radon anomaly before a March 17, 1976 earthquake (M_L = 4.3) 25 km east of Hollister. Five weeks before the event, the radon emanation began to increase significantly, reaching a maximum value 130% above the average level just before the earthquake. King[750] discovered that the anomalous behavior was not observed at two control stations 100 km northwest of the network where weather conditions were similar. Therefore, King believes that the radon anomaly was an actual precursor to the earthquake.

Birchard and Libby[752] of UCLA recently reported that for three M_L = 3.0 to 3.3 earthquakes and one 4.3 event — all centered at approximately 33° 30′N, 116° 30′W — the emanation of radon decreased prior to the quakes, with rapid increases following the shocks. They believe that the changes in radon soil gas might be a response to subsurface pore pressure variations.

Subsurface water monitoring programs have been established in three areas of California. Shortly after the August 1, 1973 Oroville earthquake (M_L = 6.0), personnel from the Laurence Berkeley Laboratory, University of California, began collecting daily water samples from several wells in the epicentral region for determining radon levels by *radiometric analysis*. In 1976, Smith et al.[753] reported that: "Some correlations have been observed between the ongoing aftershock activity and changes in radon abundance in three bedrock wells." A second program began in 1976 on the central San Andreas fault near San Juan Bautista, where a single continuous monitoring station is operated jointly by the Laurence Berkeley Laboratory and the USGS.[753]

The University of Southern California operates a ground water sampling network along the San Andreas fault from Cajon to Gorman.[754] Samples are taken from cold springs, hot springs, and water wells at weekly intervals. Teng et al.[754] report that the "data show some interesting variations when compared to the local earthquakes recorded by the Caltech (California Institute of Technology) network."

Since January 1976, the radon content of ground water has been monitored in several wells and springs in the Lake Jocassee area by personnel from the University of South Carolina.[755,756] During the first 6 months of the program, sampling was done on a weekly basis, and anomalous changes in radon emanation occurred within 36 hr of three small earthquakes (M_L's < 1.8). To better detect potential short-term changes, a continuous monitoring network (a system of electronics and *ionization chambers*) was implemented in November 1976. Since the establishment of this network, the largest earthquake (M_L = 2.3) occurred on February 23, 1977. Between February 8 and 13, the radon level dropped 70% below its normal level. This reduced level was maintained for 2 days, increasing to a near normal level 1 day before the earthquake.[756]

Mechanisms that can explain the apparent paradox of radon levels preceding certain earthquakes are unknown at this time. In reference to a diminution of radon, it could be that when rocks are compressed the release of gas is retarded.[680] However, King[751] believes that the increase in radon emanation could also be explained by the build-up of regional stress which squeezes out gases at an increased rate; the increased outgassing rate might then perturb a soil's radon concentration profile thus permitting deeper soil gases with a much higher radon content to migrate upward to the detection level. According to Sykes,[680] the increase in radon might be attributable to dilatancy, because newly formed fractures in the hypocentral regions could cause an increase in water flow with an accompanying increase in radon exhalation.

G. Ground Water Changes

Changes in the level of ground water have been reported before a number of earthquakes. Most of the changes have been observed in open wells that were located along or close to active faults. Accumulating stress might affect the level of ground water in several ways. It may (1) change the porosity of rocks causing variations in pore pressure, (2) alter ground water passageways, or (3) tilt the ground causing the flow of ground water as it seeks equilibrium.[708,757] Water level changes attributable to tectonic stress build-up can be masked by variations in precipitation, atmospheric pressure, and cultural withdrawal practices. A filtering technique can be used to remove the effects of atmospheric pressure from the raw water-level data.

Chinese accounts of changes in the water level of wells preceding earthquakes date back to 94 A.D. Between this date and 1973, 15 earthquakes were reportedly preceded by a rise or fall of up to several meters in water levels for wells sited in epicentral regions.[675] According to Bolt,[758] thousands of nonscientists are currently involved in measuring water levels in wells as part of the People's Republic of China prediction program.

An interesting example from the People's Republic of China is provided by the 1966 Hsintai earthquake. Well data are presented in Figure 14. Within the epicentral region, water levels increased by as much as 2 m, while outside this area, the level of the water dropped. The wells that displayed a premonitory change (rise or fall) were sited along the local structure line (Figure 14).[708]

In 1971, a 152-m deep well was drilled into the San Andreas fault zone to enable the continuous monitoring of water level changes and related *in situ* pore pressure changes in an area characterized by active fault creep and seismicity.[759-762] The well is located approximately 300 m north of the Almaden winery, where a concrete-lined drainage ditch and building are being sheared by fault creep episodes[759,760] (Figure 15 in Volume I, Chapter 2).

Small, but significant, water level changes have been followed by three moderate-sized earthquakes on the San Andreas fault and within about 20 km of the well: Bear Valley, M_L = 5.0, February 24, 1972; Stone Canyon, M_L = 4.7, September 4, 1972; and Lewis Ranch, M_L = 4.0, January 15, 1973. In each case, the pre-earthquake sequence of the water level was characterized by a drop followed by a rise approximating the initial value (Figure 15).[761]

As noted by Kovach et al.,[761] these observations can be explained by a dilatancy

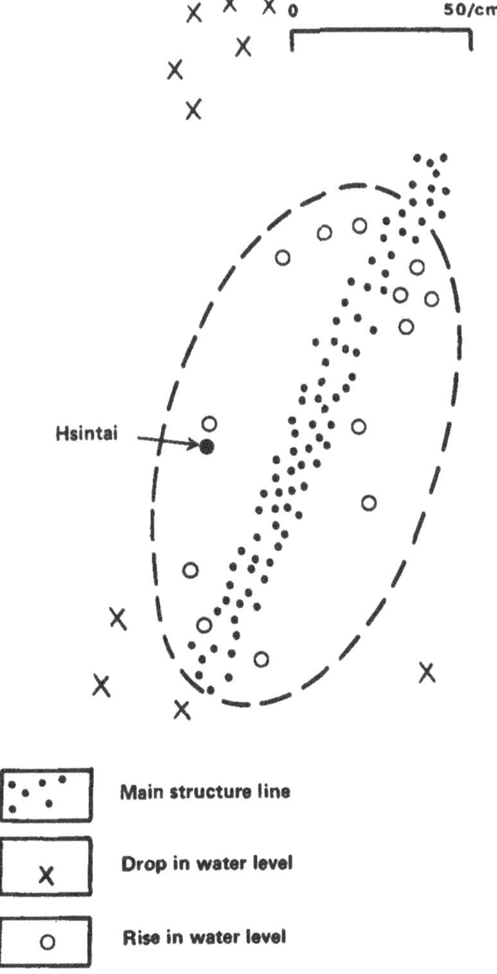

FIGURE 14. Changes in the ground water level near Hsintai, People's Republic of China before the 1966 Hsintai earthquake. (Adapted from Kuo, T.-K., Chin, P.-Y., and Feng, H.-T., *Acta Geophys. Sinica*, 17, 99, 1974 by Bolt, B.A. and Wang, C.-Y., *CRC Crit. Rev. Solid State Sci.*, 5, 140, 1975. With permission.)

process with widespread nonelastic effects and/or by a dislocation process[763,764] with nonelastic effects restricted to a narrow band along the fault. Regarding dilatancy, the initial drop in water level could be caused by the widening of existing cracks or the formation of new cracks. Water level recovery prior to the shock would be due to the inflow of ground water or to crack closure. Kovach et al.[761] suggest that additional wells are needed along the San Andreas fault to determine the actual preseismic process that is responsible for the water level changes.

Anomalous changes in the water level of this well also have preceded a number of creep episodes.[759-761] Several of the events have occurred in pairs (see days 200 to 230, 400 to 430, 550, and 930 in Figure 15). The first event of each pair is associated with a drop in the water level, while the second event is associated with a rise.

Sundaram et al.[765] have conducted laboratory shear tests on flat joint surfaces in saturated quartz monzonite to assess water pressure changes that occur during stable

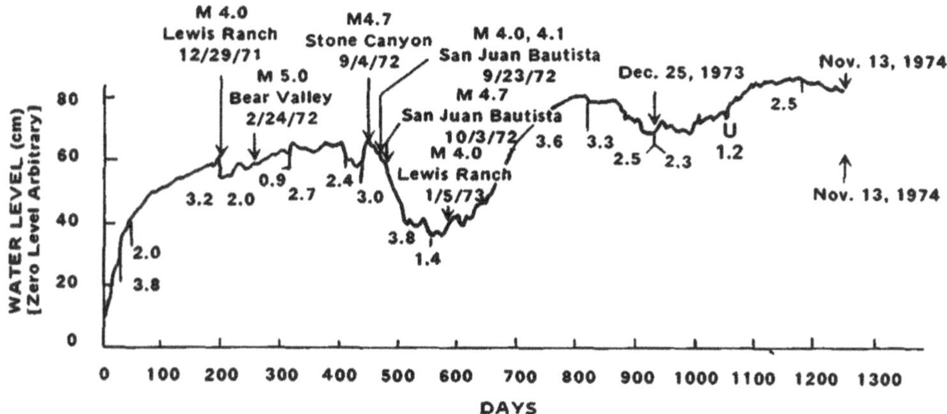

FIGURE 15. Water-level fluctuations for a well drilled into the San Andreas fault zone corrected for atmospheric pressure effects. Day 0 is May 24, 1971. The small vertical bars beneath the water-level curve indicate the time of occurrence and magnitude in millimeters of aseismic creep events near the well. Earthquakes of local magnitude 4.0 or larger that occurred on the San Andreas fault within approximately 20 km of the well are indicated on the graph. The two broad minima centered near days 550 and 950 represent the effects of a seasonal dependence on rainfall. Small but significant minima beginning on days 200, 400, and 540 were followed by earthquakes of moderate size as indicated. (From Kovach, R. L., Nur, A., Wesson, R. L., and Robinson, R., *Geology*, 3, 438, 1975. With permission of the Geological Society of America.)

sliding and stick-slip. The data indicate that the water pressure increases during post-peak shear behavior; this is most likely due to the opening of new cracks and the enlargment of existing cracks. For a *stick-slip event*, there is an accelerated drop in pressure for the stick or preseismic part and a momentary rise during the slip or coseismic part. They note that if the slip part of a stick-slip event is acceptable as a "laboratory earthquake," pressure variations observable as water level variations in wells could have "a direct bearing on earthquake prediction."

An interesting example of well water variations was associated with the 1970 Przhevalsk earthquake near Alma Ata in the Soviet Union. Sadovsky et al.[678] discovered that both the temperature and level of water in a well approximately 30 km from the epicenter increased before the event. According to Sykes,[680] more than 20 wells near Tashkent and in the Fergana Basin of Uzbekistan are being monitored for precursory changes in radon, temperature, and water level.

According to Sundaram et al.,[765] Kuo[766] reported that premonitory to several earthquakes in Taiwan, the water level in wells dropped and then rose. In some instances, changes were noted also in the color and temperature of the water.

If tectonic stress variations could be responsible for premonitory water level fluctuations in wells, might such variations also be responsible for fluctuations in the production levels of oil and gas wells? This idea has been proposed by Arieh and Merzer[767] and Wu.[768]

Arieh and Merzer noticed fluctuations in the flow of oil from wells (driven by natural pore pressure) in the Gulf of Suez premonitory to several earthquakes located close to the bifurcation point of the Gulf of Suez and Eilat (Figure 16). They proposed that there might have been preseismic crustal deformation around the sites of the impending earthquakes and that the deformation could have extended to the oil field, subsequently causing changes in pore pressure and large fluctuations in oil flow.

Wu reports on a case of pressure changes in two gas fields (Niushan and Chutouchi) in Taiwan prior to a $M_S = 6.75$ earthquake that occurred on January 18, 1964. The

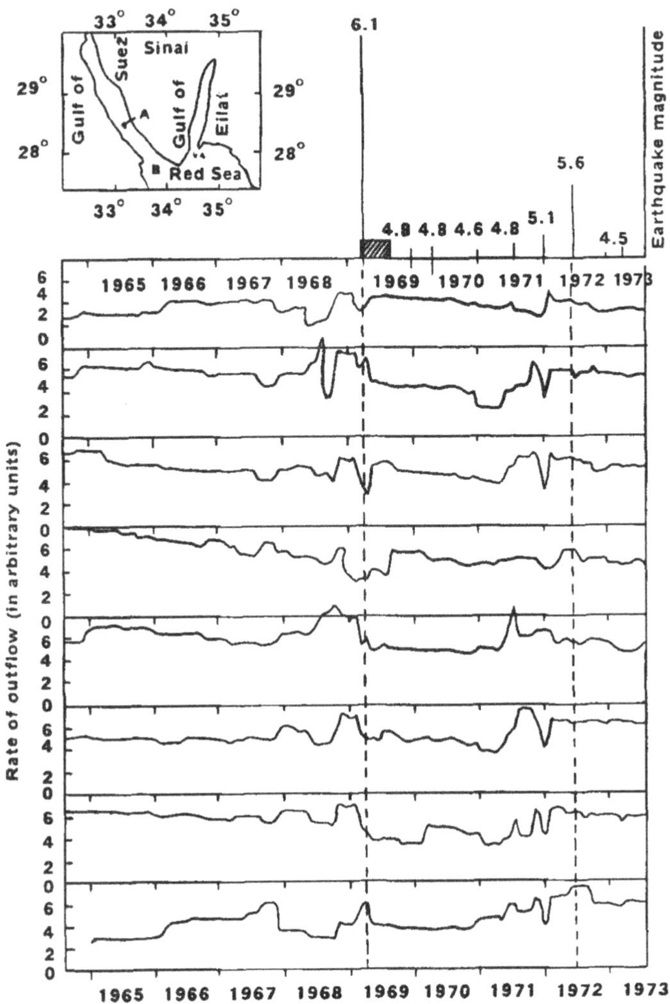

FIGURE 16. Measurements of oil flow from wells in the Gulf of Suez from July 1964 to July 1973. Earthquakes are indicated with their magnitudes. The shaded region under the 6.1 magnitude earthquake indicates foreshocks. Note the similarity of time function starting before the 1969 earthquake (magnitude 6.1) and the 1972 earthquakes (magnitudes 5.1 and 5.6). The inset is a generalized map of the Sinai region, showing the location of oil wells (A) and the location of the earthquakes (B). (From Arieh, E. and Merzer, A. M., *Nature (London)*, 247, 534, 1974. With permission.)

earthquake was probably associated with an east-dipping thrust fault as the Niushan and Chutouchi fields are within 8 and 12 km of the surface trace of the fault, respectively, i.e., within the source region of the earthquake.

Well-head pressure variations for three wells in the Niushan field are presented in Figure 17. Note the increase in pressure (both daily maxima and minima) commencing on January 8. Careful examination of records for several years before the earthquake showed no other such change. Breaks occurred in the tubing during the earthquake, causing drops in well pressure; two wells were repaired on January 20, but the third was permanently damaged (Figure 17).

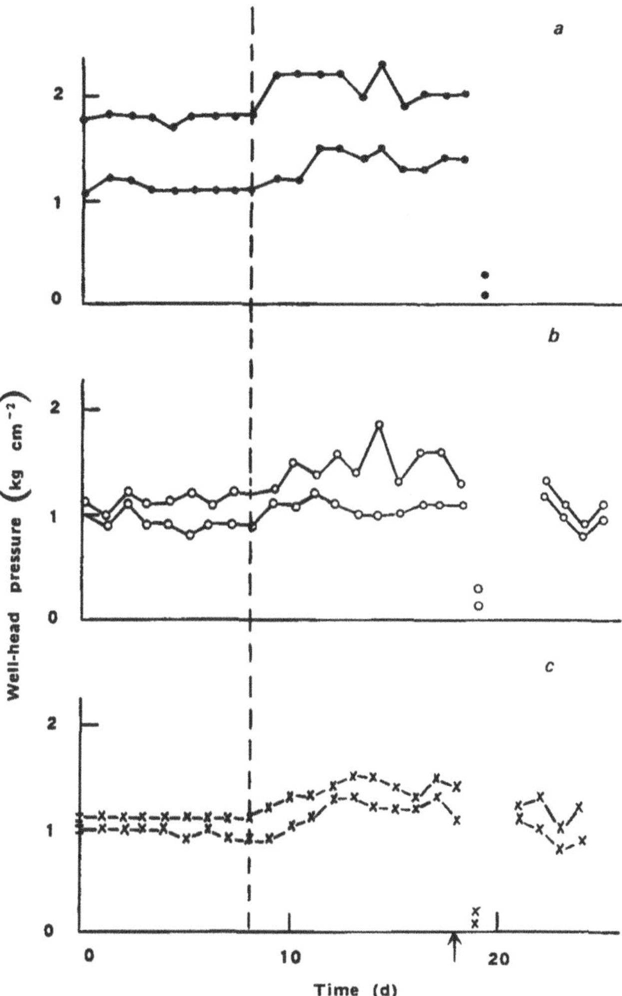

FIGURE 17. Daily maxima and minima well-head pressure variations for three wells (A,B,C) before and after the M_s = 6.75 earthquake of January 18, 1964 (arrow) at the Niushan gas field, Taiwan. The pressure started increasing on January 8 (dashed vertical line). (From Wu, F. T., *Nature (London)*, 257, 622, 1975. With permission.)

Gas production data for four wells in the Chutouchi field showed only slight changes for the period from January 1 to 18. However, upon examining production data for the wells from 1958, it was discovered that a more rapid drop commencing in March 1963 was superimposed on a long-term decrease (Figure 18). As opposed to the short-term nature of the Niushan pressure changes, the Chutouchi data are best categorized as a possible long-term precursor.

The difference in behavior of the two fields may be explained by the reservoirs being on opposite sides of the thrust fault; the Niushan field is on the footwall side and the Chutouchi field on the hanging wall side. As noted by Wu, the mechanism causing this behavior should become more apparent once the stress field around the fault is known in detail. Wu points out that because pressure and production data are continuously recorded by oil and gas companies, the records could be routinely examined for possible precursor evidence from fields located in seismically active regions.

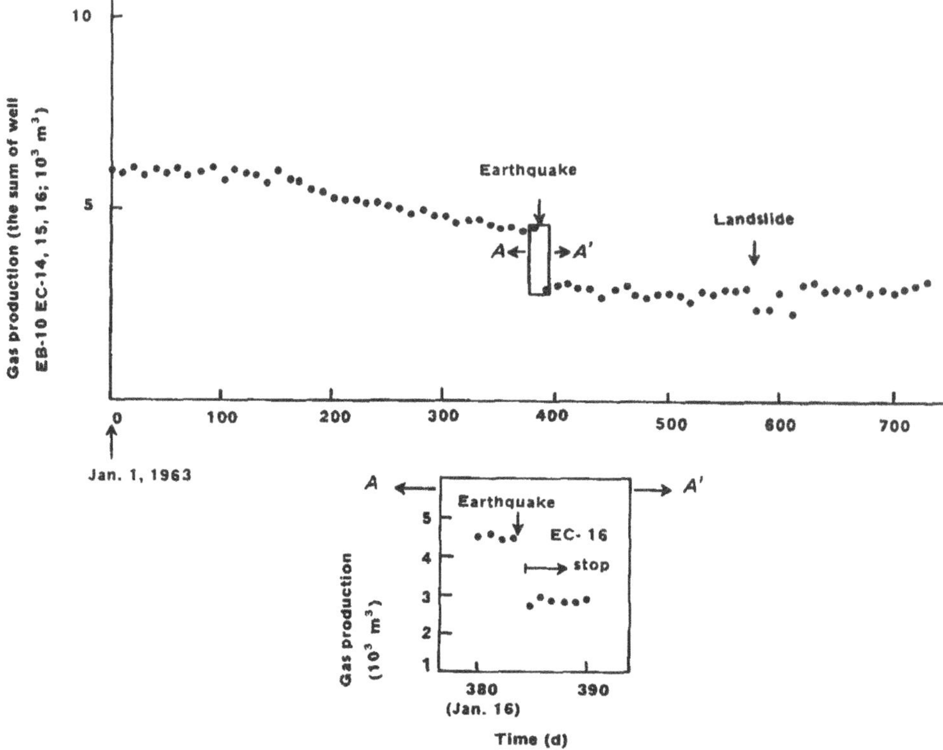

FIGURE 18. Gas production-rate data for four wells in the Chutouchi gas field, Taiwan, before and after the January 18, 1964 earthquake (M_s = 6.75). Note the more rapid drop in production beginning in March 1963. The drop in total production after the earthquake was caused by a coseismic disruption of one well (inset). (From Wu, F. T., *Nature (London)*, 257, 663, 1975. With permission.)

H. Seismic Wave Anomalies

Large-scale changes in seismic wave velocities (and hence travel times) have been detected before a number of small- and moderate-sized earthquakes. Basically, the existence of this precursor is based on the premise that if seismic waves originating from known sources propagate through or within a pending source region, velocity changes might occur as changes occur in the physical properties of a focal zone undergoing strain deformation.

Two methods are used to search for changes in seismic wave velocities. *Wadati diagrams**[769] (after the Japanese seismologist K. Wadati) are used to determine potential changes in the *travel/time ratio of secondary to primary waves* (t_s/t_p), or alternately in the *velocity ratio of primary to secondary waves* (V_p/V_s). The difference in the arrival times of S- and P-waves (t_s-t_p) for each selected earthquake (i.e., those with clear phases) is plotted against the arrival time of P (t_p) at more than two stations recording the event. The slope of the resulting near-linear line drawn through the data plots represents t_s/t_p-1, which immediately renders t_s/t_p and hence V_p/V_s.[708] The seismic sources for determining arrival time differences are usually small events located near or within (e.g., foreshocks) an expected source region for a primary earthquake.

The second method makes use of *P-wave travel/time residuals*. Based upon the description of Bolt and Wang,[708] consider the arrival time (T) of a P-wave at a station

* Wadati diagrams have been used since the 1930s to estimate the origin time of earthquakes with only limited data.

that originated at some origin time (T_o) and traveled a distance Δ to the station. The observed travel time is, therefore, $T-T_o = t_o$. If the theoretical travel time (t_c) for distance Δ is computed from average travel time tables, t_o-t_c is the residual r. Residuals can be calculated for both teleseismic and local events.

A travel/time residual is the sum of several factors, including imperfections of source determinations and velocity changes along the travel path due to an inhomogenious crust. The part of the residual due to these factors must be removed, at least to a first order, if variations in a travel/time residual due to velocity changes near a station are to be identified. Procedures for removing these effects are described by Bolt and Wang.[708]

During the 1960s, Kondratenko and Nersesov,[770] Nersevov et al.,[771] and Semenov[772] reported on apparent travel/time anomalies preceding several historical earthquakes in the Garm region, Tadzhik Soviet Socialist Republic. The anomalies were discovered ex post facto by analyzing seismograms for nearby earthquakes that were recorded at stations located in what were to become source regions for approaching events. The earthquakes were associated with thrust faulting and had magnitudes ranging from 3.0 to 5.0.

V_p/V_s ratios premonitory to the Garm earthquakes appeared to have undergone three states. The sequence commenced with a stable ratio of approximately 1.75. This stage was followed by a diminution in the ratio (approximately 10%) for 1 to 2 months; this pattern is now referred to as the *embayment*. The sequence culminated with the ratio returning to a near normal value. Earthquakes occurred shortly after this stage was reached, and an entire cycle was completed in less than 3 months. The anomalous activity in the ratio was attributed to velocity changes in the S-phase.

Semenov[772] reported that the magnitude for the Garm events appeared to be scaled with the time duration of the V_p/V_s minimum rather than to the amount of drop. Simply stated, the longer the V_p/V_s minimum persisted, the larger the magnitude of the ensuing earthquake.

The first non-Soviet observation of the velocity ratio anomaly occurred at Blue Mountain Lake (BML) in the Adirondack Mountains of New York.[773-775] A group of seismologists from the Lamont-Doherty Geological Observatory, Columbia University, had deployed a series of portable seismographs in the BML area to monitor a swarm sequence that had commenced in May 1971. Ex post facto analysis by Aggarwal et al.[773,774] and Aggarwal and Sykes[775] revealed that significant variations in V_p/V_s had occurred prior to several of the larger earthquakes associated with thrust faulting.

Days before a larger event, there was a decrease in V_p/V_s of up to 13%, which subsequently returned to a near-normal value just before the earthquakes — identical stages were found premonitory to the Garm events. No embayments were noted during stable seismic periods or when earthquake magnitudes (m_b) were less than 1.0. Similar to Soviet findings, earthquake magnitude was proportional to the time duration of the anomaly. The variations in V_p/V_s were initially linked to significant changes in V_s,[773,774] but Richards and Aggarwal[776] reevaluated the 1971 BLM data (determined V_p and V_s separately) and concluded that prior to the larger 1971 events, V_p and V_s had decreased 15 to 20% and 5 to 10% respectively, below normal values.

Figure 19 depicts V_p/V_s ratios as a function of time for the BLM area for July and August 1971. From July 1 to 3 and 11 to 12 and August 25 to 31, no events larger than $m_b = 1.0$ occurred, and the velocity ratio remained at about 1.75. Note, however, that the pronounced V_p/V_s embayments preceded the larger earthquakes on July 10 ($m_b = 3.3$) and July 27 ($m_b = 2.5$) by several days, while the shorter embayments (i.e., a few hours) preceded earthquakes on July 8, 9, and 27.

Whitcomb et al.[777] of the California Institute of Technology (CIT) were the first to

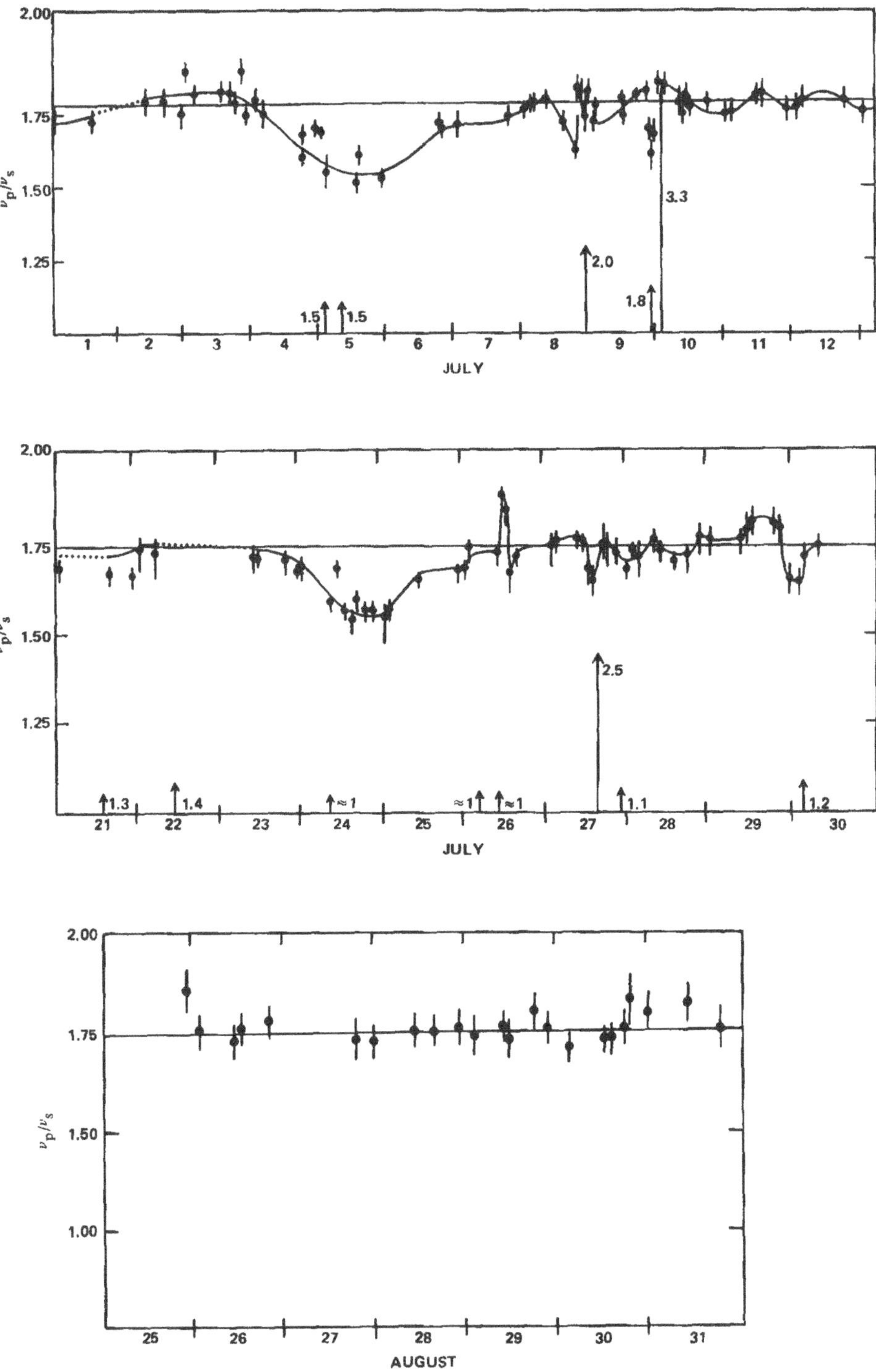

FIGURE 19. Velocity ratios of primary to secondary waves (V_p/V_s) and earthquake occurrences as a function of time for the Blue Mountain Lake area, New York for July and August 1971. Arrows and numbers indicate the time of occurrence of earthquakes and their respective magnitudes. Bars represent estimated range errors in the data. (From Aggarwal, Y. P., Sykes, L. R., Armbruster, J., and Sbar, M. L., *Nature (London)*, 241, 102, 1973. With permission.)

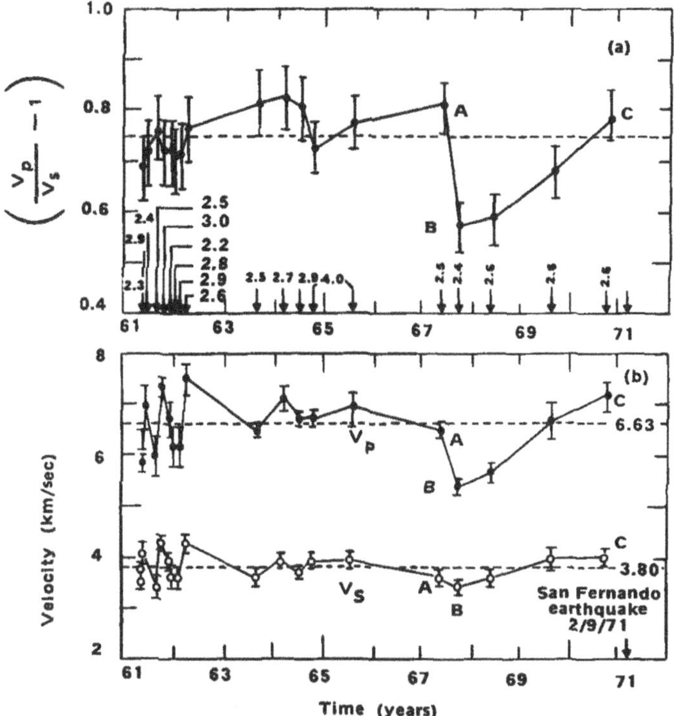

FIGURE 20. Variation of (A) $V_p/V_s - 1$ and (B) V_p and V_s between Pasadena and Riverside seismograph stations from 1961 through 1970 before the San Fernando, California earthquake (M_s = 6.5) of February 9, 1971. (From Whitcomb, J. H., Garmany, J., and Anderson, D. L., *Science*, 180, 632, 1973. Copyright 1973 by the American Association for the Advancement of Science. With permission.)

search for velocity changes preceding an earthquake with a magnitude over 5.0. Following the February 9, 1971 San Fernando earthquake (M_S = 6.5), Whitcomb et al. used seismograms recorded at CIT stations in Riverside and Pasadena (Δ = 76 km) for 19 earthquakes (M_L = 2.5 to 4.0) that had occurred to the north of both stations (i.e., close to the 1971 San Fernando epicentral region.) These events were selected because ray paths were in line with both stations. Travel times for P- and S-waves were then computed for the crustal zone between Riverside and Pasadena. It was discovered that in mid-1967 a sudden and significant decrease had occurred in V_p/V_s (about 10% below a normal value of 1.75) which slowly recovered to a near-normal value just prior to the earthquake (Figure 20). The time span between the onset of the anomaly and the earthquake was approximately 3.5 years.

According to Whitcomb et al.,[777] a velocity variation in P was largely responsible for the drop in V_p/V_s (Figure 20). Computations from travel/time data indicated that during the precursory period, V_p varied from −19 to +8% of the average velocity of 6.63 km/sec. By contrast, V_s changed from −10 to +5% from an average of 3.8 km/sec. This marked the first time that the ratio diminution was linked to V_p and not V_s.

The large variations in P-wave velocity were attributed to dilatancy occurring in water-saturated rocks comprising the focal region. Whitcomb et al.[777] postulated that as cracks opened in the strained rocks the pore pressure would drop because of increased pore volume. This state of undersaturation would cause a decrease in the rock's bulk modulus and, therefore, a more pronounced reduction in V_p than in V_s (see Formulas 1 and 2 in Volume I, Chapter 2). If the cracks subsequently fill with liquid from ground water inflow, P-wave velocity would again increase, but S-wave velocity, by comparison, would remain essentially unchanged whether the cracks were filled with air or water.

Whitcomb et al.[777] also plotted anomalous velocity time intervals as a function of magnitude for the Garm, Blue Mountain Lake, and San Fernando earthquakes. The data plots indicated the following relationship between magnitude (M) and anomally time in days (t):

$$\log t = 0.68M - 1.31 \qquad (1)$$

If this empirical relation can be extended to larger earthquakes, the precursor times for magnitude 7.0, 8.0, and 8.5 events would be 7.72, 36.96, and 80.86 years, respectively.

Successful observations at Garm, Blue Mountain Lake, and San Fernando have encouraged earth scientists to search for preseismic velocity anomalies (most often from historical records) in many seismically active regions and a sizeable number of positive and negative studies have been published. Generally, the strongest anomalies have been associated with thrust faulting. Encouraging findings have been reported for earthquakes in Taiwan,[778] New Zealand,[779,780] Japan,[781-785] the Soviet Union,[710,786,787] the People's Republic of China,[675] and the U.S. (Washington,[788] Alaska,[789] New York,[790,791] South Carolina,[792-795] California,[796-801] and Nevada[802]), while inconclusive or negative results have been reported for earthquakes in Japan,[781,784] Hawaii,[803] and California.[804-813] A sampling of these studies has been selected for discussion.

By analyzing approximately 3000 teleseismic P-wave residuals recorded at the Matsushiro Seismic Observatory in Japan from 1960 through 1968, Wyss and Holcomb[782] discovered, ex post facto, that P-wave velocities began to decrease by 20% 3 years before the beginning of the 1965 to 1967 Matsushiro swarm. The residuals returned to a normal level 330 days before the seismic activity commenced and 570 days before the principal release of energy occurred for the swarm. No long-term residual changes were noted during or following the swarm sequence. According to Wyss and Holcomb, this represented the first instance in which velocity changes were associated with strike-slip faulting.

During a 1974 visit to the People's Republic of China, Sykes and Raleigh[675] were shown 16 examples (i.e., 10 from the Peking seismic network, 5 from the Lanchow seismic network, and 1 from the Red Mountain seismic network) of premonitory changes in V_p/V_s that occurred prior to earthquakes having magnitudes between 3.5 and 6.3. In the Peking and Lanchow areas, no false alarms have occurred for events with magnitudes exceeding 3.5 and 5.0, respectively. Most of the earthquakes associated with premonitory V_p/V_s anomalies have been associated with strike-slip faulting.

The first successful earthquake prediction in the U.S., and the first in the world using travel/time anomalies, was made at Blue Mountain Lake in August 1973.[790,791] In mid-July, following two m_b = 3.6 earthquakes, seven portable seismographs were installed in the epicentral region by Lamont-Doherty seismologists. Seismograms were analyzed in the field daily to monitor for variations in t_s/t_p. For about 2 weeks, the travel/time ratio remained at a mean level of 1.73, but on July 30, t_s/t_p dropped to 1.5 and remained at this level for the next 2 to 3 days. V_p decreased approximately 22% from a normal value of 5.9 to 4.6 km/sec, and V_s dropped 12% from 3.4 to 3.0 km/sec. This pattern was similar to that observed for several of the 1971 Blue Mountain Lake (BML) earthquakes.

On the night of August 1, a prediction was made that a magnitude 2.5 to 3.0 event

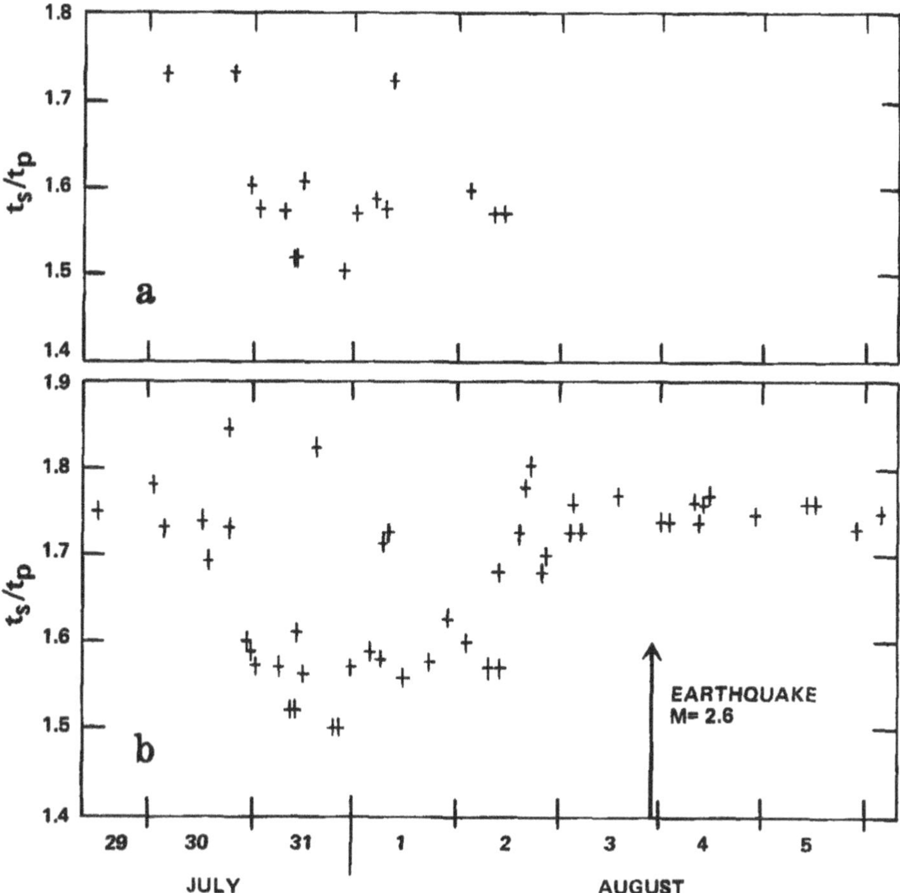

FIGURE 21. Travel/time ratios of secondary to primary waves (t_s/t_p) as a function of time before and after the occurrence of the predicted earthquake of August 3, 1973 at Blue Mountain Lake, New York. Arrow indicates the time of occurrence of the earthquake. A. Data were reduced in the field before the August 3 earthquake and used to predict the event. The anomalously low t_s/t_p values observed on July 31 and August 1 were interpreted as a precurosr to an impending earthquake. B. Data in part A combined with the results of data reduction after the occurrence of the earthquake are shown. Note that, in B, t_s/t_p apparently returned to normal before the earthquake. (From Aggarwal, Y. P., Sykes, L. R., Simpson, D. W., and Richards, P. G., *J. Geophys. Res.*, 80, 722, 1975. Copyrighted by American Geophysical Union. With permission.)

would strike the same epicentral region within the next few days. The size of the predicted event and its time of occurrence were inferred from the observed spatial extent of the t_s/t_p anomaly before it returned to its normal value. A m_b = 2.6 earthquake occurred on August 3, 1973. Travel/time ratios preceding and following this earthquake are presented in Figure 21.

In addition to monitoring local swarm activity to determine temporal t_s/t_p values, a number of distant quarry explosions were recorded at BML from a variety of azimuths before the August 3 earthquake. P-wave arrivals from these explosions showed late arrivals (i.e., P-wave velocity decreased) at five of the portable seismograph stations during the time of the premonitory low in the t_s/t_p ratio. Maximum P-wave delays of 0.13 sec (positive residuals) occurred in the hypocentral region of the pending earthquake; the amount of delay decreased along two vectors with increasing distance from the hypocentral region. Aggarwal et al.[791] stated that this spatial distribution of P-wave residuals indicated that changes in t_s/t_p were caused by changes in the physical properties of the source region.

Robinson et al.[797,798] have demonstrated the existence of P-wave velocity changes preceding three moderate earthquakes along the central San Andreas fault. For example, P-wave travel time residuals for small (M_L = 0.7 to 3.6) local (Δ 20 to 70 km) earthquakes recorded at a USGS seismograph station displayed significant variations before the Bear Valley earthquake (M_L = 5.0) of February 24, 1972. The station (BVL) was located approximately 2 km from the epicenter. Residuals increased nearly 0.3 sec above normal values (i.e., P-wave velocity decreased) approximately 2 months before the event but returned to a normal level several weeks in advance of the earthquake. With two exceptions, residuals at two nearby stations did not exceed the normal range of ±0.15 sec. According to Robinson et al.[797] the amount of change in BVL residuals "could be explained by a 10 to 15% decrease in P-wave velocity within a volume of the same radius as the observed aftershock zone (7 km)."

Cramer[799] also identified P-wave velocity changes premonitory to the 1972 Bear Valley earthquake. However, he used teleseismic P-wave residuals for large earthquakes occurring in the Circum-Pacific seismic zone. Teleseismic arrivals for the period from July 1971 through April 1972 were analyzed by the two-station residual method (described by Bolt and Wang[708]) and corrected for azimuthal variations. During a part of January 1972, a 0.15 sec vertical travel/time delay occurred beneath the BVL station, while the P-wave travel time decreased by 0.2 sec (i.e., P-wave velocity increased) beneath a second station located 13 km from the epicenter. According to Cramer, these two-station data "support a limited radial extent of five to ten kilometers for P-velocity delays associated with magnitude 5 events along the San Andreas fault in central California with increased P-velocity sometimes occurring at greater distances."

Conflicting data have been presented for the same area in central California.[804,808] By using teleseismic arrivals for earthquakes occurring in the Fiji-Tonga-Kermadec Island source region and recorded at four stations of the University of California, Berkeley, seismographic network, Cramer and Kovach[808] found no premonitory P-wave velocity variations at stations closest to the 1972 Bear Valley and October 3, 1972 San Juan Bautista (M_L = 4.9) earthquakes. The station-epicenter distances were 23 and 11 km, respectively. However, a small anomalous P-wave velocity zone could have gone undetected because of these distances and, hence, unfavorable wave paths. Cramer and Kovach believe that if V_p anomalies do occur along the San Andreas fault their data would indicate a local velocity zone with a radius of about 10 km for magnitude 5 events.

Using travel times of P- and S-waves from quarry blasts that were measured at seven stations of the University of California, Berkeley, network for the period from mid-July 1961 to mid-June 1973, McEvilly and Johnson,[804] found no correlative evidence of velocity changes preceding moderate earthquakes along the central San Andreas fault. Velocities were generally within ±1% of the normal level, with occasional deviations amounting to ±2%. According to McEvilly and Johnson, these variations were most likely accountable to reading errors and source location changes at the quarry.

Although tilt[693] and magnetic[740] anomalies were observed before the November 28, 1974 Hollister earthquake, it is unclear whether the earthquake was preceded by a P-wave velocity anomaly. Lee and Healy[801] reported that a P-wave delay of 0.2 sec was found at three USGS seismograph stations near the epicenter. The anomaly commenced 1.5 months before the earthquake and persisted for about 1 month. The travel/time residuals were determined from events occurring in the Bear Valley region, approximately 35 km south of Hollister.

Cramer[810] could find no preseismic travel time delays exceeding 0.1 sec for telese-

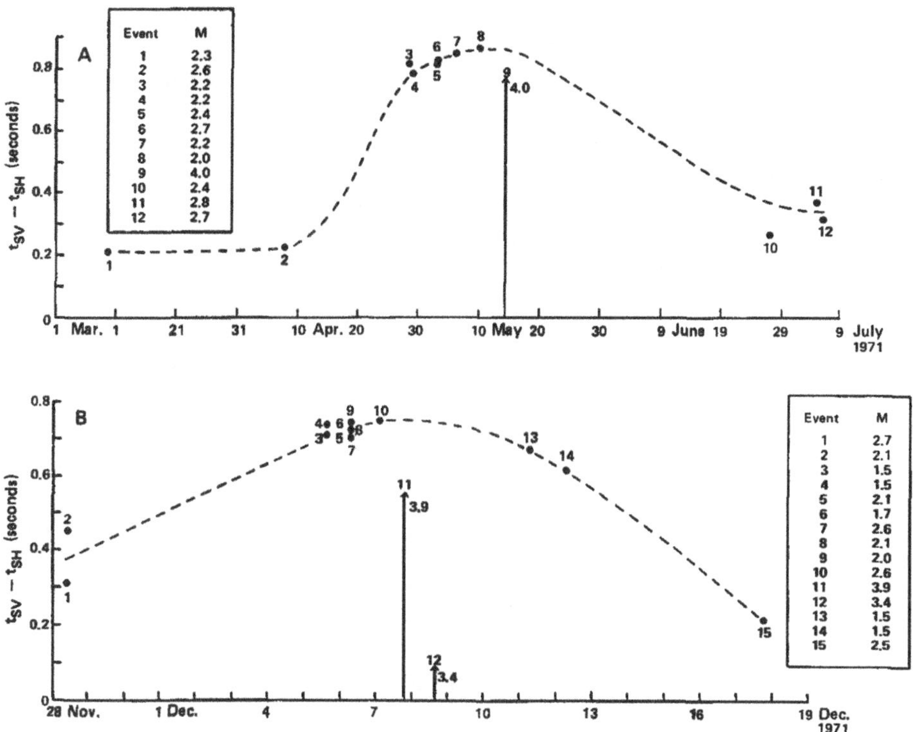

FIGURE 22. Observed temporal variations in S-wave velocity anisotropy ($\Delta t = t_{sv} - t_{sh}$) for earthquakes in (A) the Slate Mountain region and (B) the Mina region, Nevada. The maximum difference in the SH and SV velocities in the two regions amounts to only 2.3 and 2.5%, respectively. The insets show magnitude for each event. The S-waves from the largest events (Event 9 from Slate Mountain and Events 11 and 12 from Mina) could not be examined because of signal clipping. (From Gupta, I. N., *Science*, 182, 1130, 1973. Copyright 1973 by the American Association for the Advancement of Science. With permission.)

ismic arrivals from Circum-Pacific source events at 30 USGS seismograph stations in the Hollister-Bear Valley region. Residuals for the station closest to the hypocenter (2.8 km) showed no premonitory time changes, suggesting to Cramer "that no anomalous P-velocity zone existed in the hypocentral region prior to the Thanksgiving Day event."

Gupta[802] has reported on a different type of seismic wave velocity anomaly. Prior to the occurrence of two 1971 earthquakes ($M_L = 4.0, 3.9$) in the Slate Mountain and Mina regions of central Nevada, Gupta observed progressively longer time separations between the two S-wave components (*S-wave velocity anisotropy*) from a number of local earthquakes (Figure 22). The earthquakes had epicenters within 10 km of the two main events, and all shocks were recorded at a three-component seismograph station in Tonopah.

Figure 22 shows the temporal variations in S-wave velocity anistotropy (Δt) for the Slate Mountain and Mina regions. Note that in both examples, Δt steadily increased with the primary shocks occurring shortly after the peak value of Δt was reached. Following the main events, Δt gradually returned to a near-normal value. Gupta associated the preseismic increases in Δt with strain accumulation. This was in agreement with the laboratory and theoretical studies of Nur and Simmons[814] and Nur.[815] They demonstrated that the application of nonhydrostatic stress to a rock specimen containing cracks created S-wave velocity anisotropy, which increased with increasing stress.

Based upon the results in Nevada, Gupta[802] proposed that (1) it might be possible to monitor crustal stress changes by systematically observing Δt and (2) its peaking at a high value might be a signal for a pending earthquake. Also, because an area can be monitored by a single three-component seismograph and results are essentially free of local effects, Gupta stated that "this method has the potential for being one of the simplest ways of predicting earthquakes."

In 1976, James Whitcomb[816] of CIT reported on a V_p/V_s embayment for an area with dimensions of about 100 km in the Transverse Ranges of southern California that approximates the epicentral region of the 1971 San Fernando earthquake. A mathematical filtering technique was used to "smooth" the V_p/V_s data. This technique is designed to eliminate embayments associated with small shocks while retaining those that might precede moderate-sized events. Smoothed V_p/V_s data began to decrease in early 1974 and remained diminished for about 1 year. The data overshot the normal level in December 1975. According to Whitcomb's embayment hypothesis, the 1.8 year diminution called for a M_L = 5.5 to 6.5 earthquake with a 1-year time window (April 1976 to April 1977).

The above data were presented at the annual meeting of the American Geophysical Union and at a CIT news conference in April 1976. The latter received coverage from the California and national news media. For example, several articles were published by the *Los Angeles Times*.[817-820] The Whitcomb pronouncement was often misinterpreted as an official prediction, and a councilman from Los Angeles urged that legal action be initiated for the alleged harm that the "prediction" would have on property values in the San Fernando Valley.[819]

Although a M_L = 4.7 earthquake occured in April 1976, its size was well below that required by an 1.8 year embayment duration. V_p/V_s returned to low levels after the event, but the required earthquake did not occur. The hypothesis test was concluded in December 1976 with negative results.[813]

I. Anomalous Animal Behavior

There are many published accounts, some dating back centuries, of *anomalous animal behavior* predating (days to seconds) the occurrence of earthquakes. In fact, Pliny advocated in 100 A.D. that such behavior be used for predicting earthquakes.[821] Anderson[822] has conducted an extensive literature search for historical accounts of different types of anomalous animal behavior that reportedly predated seismic events. Simon[823] has summarized these findings:

> Zoo animals refuse to go into their shelters at night; snakes, lizards, and small mammals evacuate their underground burrows; hyperactive insects congregate in huge swarms near seashores; cattle seek high ground; wild fowl leave their usual habitats; domestic animals become agitated.

The earliest published accounts in the U.S. were for the 1906 San Francisco earthquake. Miss Finette Locke kept detailed notes of unusual animal behavior cases that were reported to her. A summary of her notes was published in the 1908 report of the State Earthquake Investigation Commission; several of the observations were:[517]

> Horses whinnied before the shock . . . Several instances were reported where cows stampeded before the shock was felt by the observer. In other cases cows about to be milked are said to have been restless before the shock . . . Lowing and bellowing of the cattle at the time of the shock was very commonly reported, and in some cases this is said to have occurred a little before the shock. The most common report regarding the behavior of dogs was their howling during the night preceding the earthquake.

Until quite recently, accounts like these were usually met with skepticism. However, this view is changing, largely due to the apparent successes the Chinese have had in using erratic animal behavior as an earthquake precursor. For example, in the People's Republic of China, farmers are instructed to watch for unusual activity in their animals, and observers are even stationed in the Peking Zoo to watch for any unusual animal activity. Erratic behavior is reported to a local seismological brigade.[675]

In 1973, the Seismologial Office of Tientsin issued a six-page pictorial booklet to observers that describes several aspects of using anomalous animal behavior for predicting earthquakes. Based upon a translation by W. H. K. Lee of the USGS, the booklet states:[824]

It is easy and simple to use animals to predict earthquakes. Certain organs of animals may acutely detect various underground changes before earthquakes. Both historical and recent surveys of large earthquakes prove that animals have precursory reactions.

This is followed by an earthquake prediction verse:[824]

Animals are aware of precursors before earthquakes; Let us summarize their anomalous behavior for prediction. Cattle, sheep, mules, and horses do not enter corrals, Rats move their homes and flee. Hibernating snakes leave their burrows early, Frightened pigeons continuously fly and do not return to nests. Rabbits raise their ears, jump aimlessly and bump things, Fish are frightened, jump above water surface. Every family and every household joins in observation, The people's war against earthquakes must be won.

During a 1974 visit to the People's Republic, Sykes and Raleigh[675] were told that unusual animal behavior was used as one of the principal precursors for predicting earthquakes in advance of their occurrence. The abnormal behavior was apparently used to localize an event in time and space.[675]

The members of the Haicheng Earthquake Study Delegation[825] who visited the People's Republic in 1976 were informed that frequent reports of anomalous animal behavior were received at local seismology offices prior to the February 4, 1975 Liaoning Province earthquake (M_L = 7.3). Throughout the pending epicentral region, the frequency of reports increased from December 1974 to the time of the earthquake. The delegation was unable, however, to determine the importance given to these accounts in formulating the official prediction. It was also discovered that earthquake-animal behavior research was being conducted at the Institute of Biophysics in Peking and at Peking University.[825]

The growing interest in anomalous animal behavior in the U.S. is perhaps mirrored best by the fact that the USGS sponsored a 2-day conference on the subject in October 1976.[826] Some 35 presentations were made by seismologists, geologists, and biologists. There was agreement that "there may be some truth in the belief that animals can sense some environmental change that precedes an earthquake."[824]

At least two government-sponsored animal studies are currently being conducted in California. The activity of captive pocket mice and kangaroo rats is being monitored at sites near the Palmdale bulge [821,824] (Figure 11), and the motor activity of cockroaches is being monitored at sites near Hollister, Twin Lakes, and Anza — sites close to active faults.[827,828] Preliminary results from the second study indicate that before the occurrence of small earthquakes, there is a marked increase in their motor activity.[828]

Scientists are also beginning to conduct interviews in areas struck by earthquakes for possible accounts of unusual animal behavior. For example, Barry Raleigh[829] of the USGS interviewed several housholds owning animals following the November 28, 1974 earthquake (M_L = 5.2) that occurred about 16 km north of Hollister, California. Out of eight interviews, one proved to be positive. Bobbi Munson, who operates a horse ranch, was attending to several horses on the day of the shock. She observed that the normally calm-natured animals displayed abnormal behavior during the morning of November 28 (the earthquake occurred at 3:01 p.m.). For instance, one horse refused to eat; a colt could not be handled; other horses ran around erratically and one fell. According to Raleigh, Ms. Munson was the only one of the eight dealing directly with animals on the day of the earthquake. This earthquake was preceded by tilt[693] and magnetic[740] anomalies.

Although some cases of anomalous behavior may be accountable to animals sensing vibrations from very small foreshocks that go undetected by humans in the same area, it is not known what animals might possibly sense as strain accumulates before an earthquake. As noted by Logan,[821] the behavior anomaly is not restricted to a single species or genus, which suggests that there are either multiple precursive stimuli or a single stimulus that can be sensed by many different types of animals.

This subject was discussed at length at the 1976 USGS conference, and several possiblities were presented.[821]

1. Changes in water level might cause certain abnormal reactions, such as snakes moving to the surface during the hybernation period.
2. Certain animals (e.g., dogs) might sense acoustic emissions from the microfracturing of rock if the activity extended to the near surface and within close proximity to the animals.
3. Although they have not been investigated in detail, changes in electrostatic effects or gas emissions may be possible stimuli.

J. Multiple Precursor Observations

The simultaneous observation of multiple precursors is important not only for a better understanding of the physical processes that precede and accompany earthquakes but also for achieving reliability in predictions.[787] Whenever a single geophysical anomaly is observed, there is always the possibility that it could be due to random occurrence. However, in qualitative terms, as the number of precursor observations increases, the probability of a false alarm decreases.

Because most seismic regions are not covered by networks of instruments that permit the simultaneous detection of geophysical anomalies, multiple precursors have been observed before only a small number of earthquakes. In the U.S., for example, two well-defined geophysical anomalies have been observed simultaneously before only one moderate-sized event — the November 28, 1974 Hollister, California earthquake (M_L = 5.2).* The crustal tilt[693] and magnetic[740] anomalies were recorded on instruments comprising a USGS prototype system operative along one of the most active segments of the San Andreas fault in central California.

The successful prediction of the February 4, 1975 Liaoning Province earthquake (M_L = 7.3) used the highest number of potential precursors yet observed before a seismic event. These were used to establish a temporal prediction sequence comprised of four stages: (1) long-term, (2) middle-term, (3) short-term, and (4) imminent. Each stage effectively narrowed the prediction window in time and space.[825] The following descriptions of each stage have been summarized from the reports of the Haicheng Earthquake Study Delegation[825] and Scholz.[830]

Long-Term Prediction — In 1970, the National Conference for Seismological Work designated the Liaoning Province as an area that might be struck by a damaging earthquake in the future because it appeared that large events were migrating northeast towards Liaoning Province (Figure 23). The temporal sequence had commenced in 1966 with two Hsingai earthquakes (M_L = 7.2 and 6.8). These events were followed

* A nongeophysical anomaly (i.e., unusual animal activity) might also have preceded this earthquake.[829]

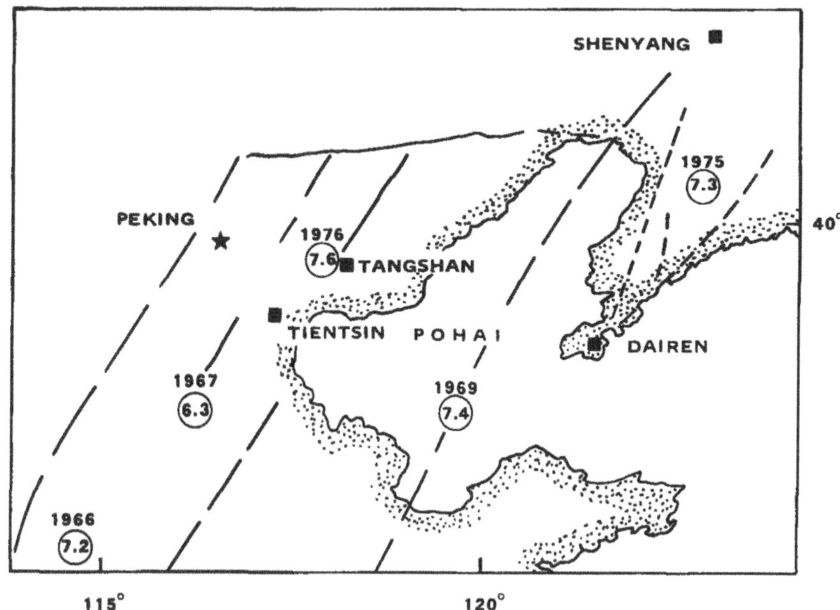

FIGURE 23. Fault map of the northern region of the People's Republic of China. Large earthquakes ($M_L \geq 6.5$) since 1966 are shown as circles enclosing the magnitude value. (From Haicheng Earthquake Study Delegation, Raleigh, B., et. al., *EOS Trans. Am. Geophys. Union*, 58, 240, 1977. Copyrighted by American Geophysical Union. With permission.)

by the 1967 Hochien (M_L = 6.3) and the 1969 Pohai Gulf (M_L = 7.4) earthquakes (Figure 23). As a result of the long-term prediction, a number of geological, seismological, and geophysical studies were initiated in the province. Some of the efforts included: (1) mapping of major faults, (2) investigation of deep fault structure by magnetic, gravity, and explosion techniques, (3) installation of 17 new seismic stations, (4) releveling surveys along the Pohai coast, (5) establishment of short leveling lines across the prominent NNE trending faults, (6) installation of tiltmeters at the Shenyang and Shihpengyu geophysical observatories, and (7) initiation of geomagnetic observations at Dairen (Figure 24).

Middle-Term Prediction — In June 1974, a second conference (State Seismological Bureau Conference) was held to evaluate the data collected in the province since 1970. Several significant findings were presented:

1. Commencing in late 1973 and continuing into mid-1974, the frequency of small shocks had increased several-fold throughout the province compared to previous years.
2. A releveling survey indicated that most of the Liaotung Peninsula had been uplifted and tilted towards the northwest between 1958 and 1970 to 1971.
3. This regional deformation of the peninsula was accompanied by accelerated tilting to the west-northwest of a short, level line crossing a fault at Jinxian (Figure 24).
4. The difference in the vertical component of the magnetic field at Dairen and Peking (Figure 23) changed by more than 20 gammas between October 1973 and May 1974.

Consequently, it was predicted in June 1974 that a magnitude 5 to 6 event would occur

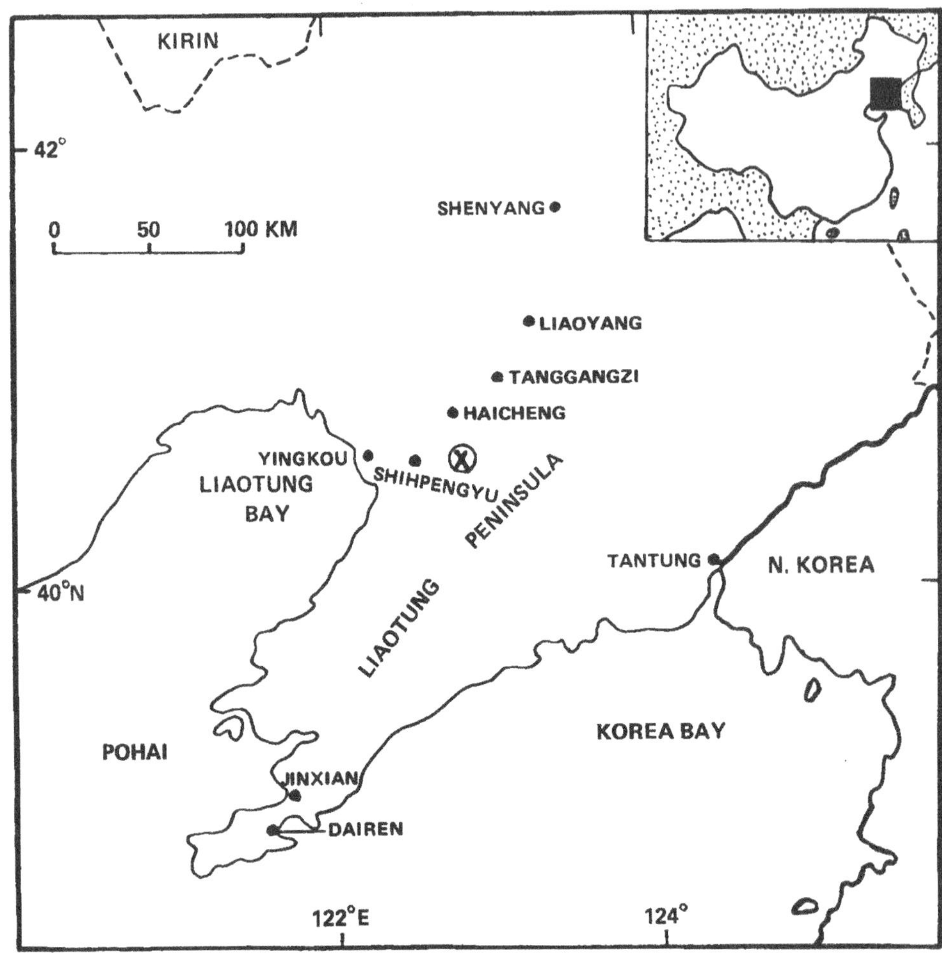

FIGURE 24. Portion of Liaoning Province, People's Republic of China, and the epicenter (⊗) of the February 4, 1975 Haicheng earthquake ($M_L = 7.3$). Certain precursors, described in the text, were observed at annotated locales.

in the northern Pohai region in 1 to 2 years. Based upon this middle-term prediction, the Party and Revolutionary committees began a mass educational program to apprise the populace of the government's prediction efforts and disaster prevention techniques. In addition, several thousand amateur observation posts were established throughout Liaoning Province for the purpose of monitoring telluric currents, radon emanation, water well levels, magnetic inclination, and animal behavior. For example, approximately 230 stations were established in the community of Yingkou (Figure 24).

Short-Term Prediction — Between June 1974 and January 1975, several different types of anomalies were observed:

1. The Liaotung Peninsula continued its tilting to the northwest throughout the latter half of 1974.
2. Anomalies in water well levels and animal behavior were reported at Tantung.
3. Beginning in December, increases in radon emanation were reported from several locations, includingTantung and Tanggangzi.
4. A swarm sequence (largest $M_L = 4.8$) commenced on December 22 in the vicinity of Liaoyang, a region normally characterized by its low seismicity.

5. The accelerated tilting rate to the WNW at Jinxian slowed and then reversed direction in early January (Figure 24).

On January 13, 1975, it was predicted at another State Seismological Bureau Conference that a magnitude 5.5 to 6.0 earthquake would occur in Shenyang-Tantung-Dairen region (Figure 24) during the first half of 1975. The Liaoning Provincial Revolutionary Committee took immediate steps to have earthquake hazard information reach every family in the region.

Imminent Prediction — The foreshock activity* of the February 4 Haicheng earthquake commenced in the impending epicentral region on February 1 and peaked on February 3. The largest foreshock (M_L = 4.8) occurred at 8:00 a.m. on February 4. Foreshock activity essentially ceased for 6 hr before the main event. Based upon previous experience, "close successive foreshocks followed by a period of calm, and then the big shock," the quiescent period was considered to be the final precursor. According to the members of the Haicheng Delegation,[825] the foreshocks were "the most important observation defining the impending earthquake in both space and time."

On the morning of February 4, the above observations, but especially the foreshock data, were used by the Provincial Earthquake Office to reduce the spatial region to the Haicheng-Yingkou area (Figure 24), and the time window was set at 1 to 2 days. The official earthquake warning was issued by the Provincial Revolutionary Committee at 2:00 p.m. Evacuation and disaster prevention procedures were carried out during the afternoon and early evening. The earthquake struck at 7:36 p.m. on a northwest striking, left-lateral fault approximately 25 km south of Haicheng (Figure 24).

Although the Haicheng event was the first major earthquake in the world to have been predicted successfully,[825] its magnitude was underestimated. The upper limit of the expected shock had been set at magnitude 6 in the 1974 middle-term prediction, and it was not changed in either the short-term or imminent predictions. Scholz[830] believes that the conservative estimate was likely influenced by the fact that the province only had two highly damaging earthquakes (magnitudes probably exceeded 6) in 3000 years of Chinese history. According to Scholz,[830] if the seismologists would have used "pure induction," they would have set the expected magnitude at 6.5 to 7.5 — the size range of the other events in the northeast migration series (Figure 23) and would, therefore, have predicted correctly the size of the Haicheng earthquake.

III. EARTHQUAKE PRECURSOR MODELS

The discovery of a number of promising precursive phenomena from a variety of geologic environments has led to the formulation of physical models that attempt to connect these geophysical anomalies with crustal processes which might be operative in pending focal regions. To date, four different types of *earthquake precursor models* have been proposed. Rock dilatancy is a crucial component for two types of models. The third identifies premonitory fault creep as the primary mechanism, while the fourth type of model identifies a propagating deformation front as the mechanism creating earthquake precursors.

Although there are arguments lending support to each model, inadequate field data have precluded the validation of any particular model. It is possible that when such data become available, the results may identify a process not proposed in any current model.

* Identified as foreshocks because of the relatively high proportion of larger sized events as compared to smaller sized shocks associated with a swarm sequence.

A. Dilatancy Models

The term *dilatancy* was used by Reynolds[831] in 1885 to describe the volume increase in "granular masses" caused by deformation. Subsequent laboratory research by Mead,[832] Bridgman,[833] Robertson,[834] Matsushima,[835] Handin et al.,[836] Frank,[837] Brace,[838] and Brace et al.,[839] among others, has shown that rocks also increase in volume relative to elastic changes (i.e., porosity increases slightly) as they are deformed.

The findings of Brace[838] and Brace et al.[839] played an especially important role in the development of *dilatancy models*. For example, it was demonstrated that as brittle rocks (e.g., granite, marble, aplite) were subjected to triaxial compression, microcracks were formed parallel with the axis of maximum compression when the applied stress reached one third to two thirds the fracture stress at a given confining pressure. The dilatancy magnitude was generally 0.2 to 2.0 times larger than elastic changes that would have been expected if the rocks were elastic.[839]

The laboratory results of Nur and Simmons[840] were also used to establish parameters for one type of dilatancy model. Their work demonstrated that, for low porosity rocks, V_p/V_s decreased in dry rock and increased in saturated rock. This was attributed to P-wave velocity being greater in saturated as opposed to dry rock, with the S-wave velocity remaining essentially unaffected. The presence of water without pore pressure increases a rock's bulk modulus; however, the rigidity modulus remains approximately equal to that of dry rock. Therefore, P-wave velocity would be more strongly affected by the presence of water[840] (see Formulas 1 and 2 in Volume I, Chapter 2).

The dilatancy mechanism was incorporated into a water-dependent *dilatancy-diffusion model* by Nur[841] in 1972. Modified versions were later introduced by Aggarwal et al.,[774,791] Scholz et al.,[842] Whitcomb et al.,[777] and Anderson and Whitcomb.[843] Basically, these models include the following stages:

1. Strain accumulates in crustal rock along a fault zone.
2. Stress-induced microcracks open and spread or existing cracks are enlarged because of increasing deformation near the pending focal region.
3. Ground water from the surrounding area diffuses into the undersaturated dilated zone filling the cracks.
4. The presence of ground water increases the pore pressure which weakens the rock, and an earthquake occurs at a point near maximum stress.
5. Following the earthquake, the cracks close and water is forced out of the dilatant zone, a process called *dilatancy recovery.*
6. A new cycle of strain accumulation commences.

Diffusionless-dilatancy models that are independent of water migration have been developed by Myachkin et al.,[844] Stuart,[845] and Brady.[846-848] Rather than the cracks filling with fluids in the dilated zone (phase 3 above), the newly opened cracks are thought to close before the earthquake because of stress relaxation caused by either (1) the formation of one or more main fractures from an "avalanche type growth" of small cracks,[844] (2) the strongly nonlinear constitutive properties of a fault-gouge zone (some type of breccia or granular aggregate),[845] or (3) a reorientation of the principal stresses in the focal region brought about by microcrack clustering.[846] With stress drop and crack closure in the surrounding region, the stress becomes concentrated in the focal zone where rapid deformation and subsequent faulting (and the earthquake) take place. Following the earthquake, the original properties in the dilatant zone are recovered.[849]

Most of the diffusion models attempt to explain premonitory changes in V_p/V_s. For example, the Nur model[841] shows that the diminution in V_p and, hence, in V_p/V_s can be caused by the opening of new cracks in *dry rock* constituting the focal zone. Whitcomb et al.[777] and Scholz et al.[842] propose that the drop in V_p/V_s is caused by the widening of existing cracks[777] and the opening of new cracks[842] in *saturated rock,* wherein the cracks ultimately grow to such an extent that they are no longer filled with water (i.e., pore volume exceeds fluid volume) — a condition termed *undersaturation.* The models[777,841,842] agree, however, that the subsequent recovery in V_p/V_s occurs as ground water diffuses into the dilatant zone. Once the cracks reach a saturation state: (1) V_p/V_s is at a near-normal value because of the increase in P-wave velocity and (2) the increased pore pressure weakens the rock, causing ultimate failure and the earthquake. Predicted changes in V_p/V_s as a function of time during the earthquake cycle for the Scholz et al. model[842] are depicted in Figure 25.

During the stage of undersaturation in the Whitcomb et al.[777] and Scholz et al.[842] models, the lack of water in the cracks strengthens the rock comprising the dilatant zone (increase in fracture strength) and more strain can be accomodated. Hence, the occurrence time of the earthquake is delayed — a process called *dilatancy hardening.*[837]

The preseismic pattern in V_p/V_s can also be explained by the diffusionless models.[845,846] The ratio would drop during dilatancy because the cracks are dry; the return to a predilatancy value would occur during the later phases of crack closure (V_p/V_s values should be similar whether the cracks are water filled or closed). In the diffusionless models the premonitory changes are also linked to variations in P-wave velocity.

Other types of precursors are implied by the dilatancy-diffusion model of Scholz et al.[842] (Figure 25). Because electrical resistivity is dependent upon the amount of water in rocks, the model proposes that resistivity will decrease during dilatancy and the influx of ground water into the dilatant zone (Stages II and III, Figure 25). However, unlike the V_p/V_s anomaly, resistivity should continue to decrease until the time of the earthquake. With the coseismic stress drop, the cracks will close and water will be forced out of the source region, subsequently, resistivity will increase towards its predilatant value in Stage VI (Figure 25).

The rate of water flow should increase steadily in Stage II due to the increased water transport into the dilated zone. The flow rate in Stage III might vary as indicated by the dashed lines in Figure 25. With dilatancy recovery, the water flow rate would drop rapidly in Stage VI (Figure 25).

Dilatancy brings about an increase in rock volume; consequently, vertical movements, crustal tilts, and volumetric strain anomalies might be expected to be associated with some earthquakes. Rapid crustal movements would be expected during Stage II; however, there would be little movement in Stage III because dilatancy has decreasd with ground water diffusion representing the dominant process. With the coseismic stress drop and dilatancy recovery (Stages V and VI), the crustal uplift would subside (Figure 25).

Premonitory seismicity decreases during dilatancy hardening in Stage II and increases at the end of Stage III (Figure 25), an increase which is coincidental with the recovery of V_p/V_s (Figure 25). This temporal seismicity pattern would indicate a quiet period some time before the main event and a short period of increased activity just before the primary earthquake (Figure 25). According to Scholz et al.,[842] dilatancy hardening may explain why the above preshock history prevails rather than a gradual increase in seismic activity leading up to the primary earthquake.

According to these researchers, the b-value (not shown in Figure 25) or the slope of the earthquake-magnitude relation, should also be affected by dilatancy since the laboratory and theoretical studies of Scholz[850] and Wyss[851] indicate that this value is primarily a function of applied stress: the value of b decreases with increasing stress.

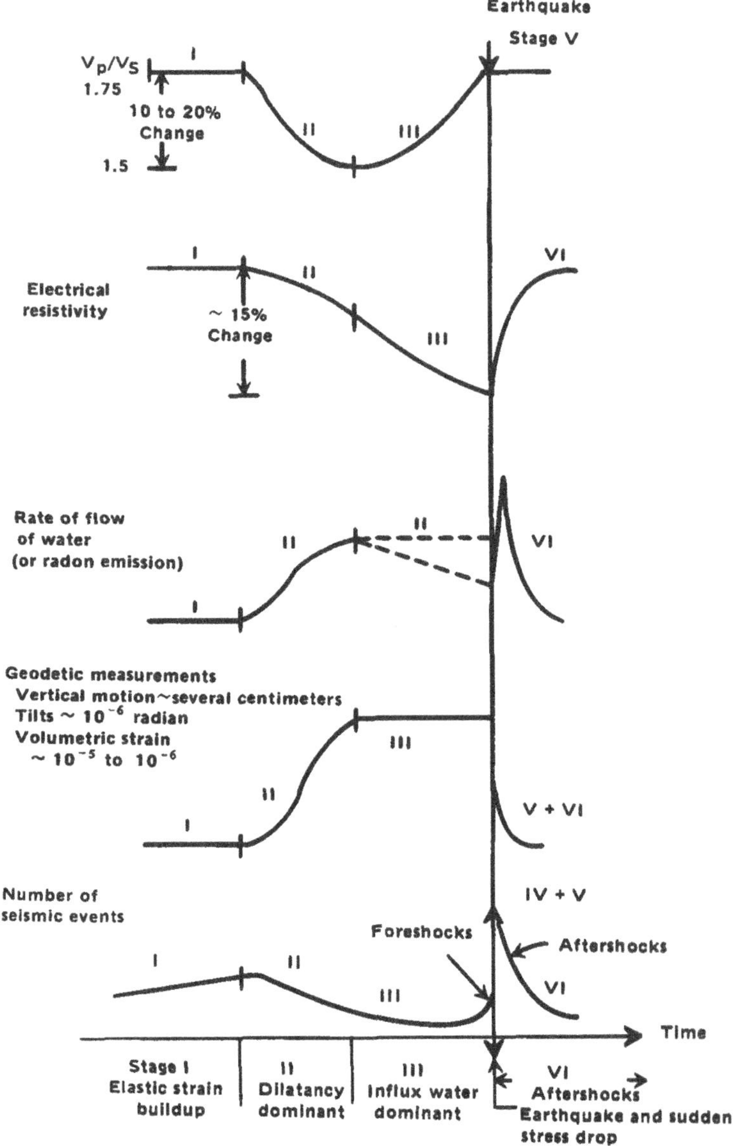

FIGURE 25. Predicted changes in various *precursors* as a function of time during the earthquake cycle for the Scholz et al.[842] *dilatancy-diffusion model.* Roman numerals indicate various stages in the cycle. Short-term fluctuations (Stage IV), which are observed before some large earthquakes, are not indicated on the sketches. The rate of water flow may vary as indicated by the dashed lines. Radon emission may be a function not only of the rate of water flow but also of the rate of creation of new surface area by the growth of cracks. (From Scholz, C. H., Sykes, L. R., and Aggarwal, Y. P., *Science,* 181, 803, 1973. Copyright 1973 by the American Association for the Advancement of Science. With permission.)

Therefore, the b-value would be expected to decrease throughout Stages II and III (dilatancy hardening), because the local stress required to cause fracturing in these stages has to be greater than prior to hardening. The b-value should increase just before the main shock.

The temporal patterns in seismic activity and electrical resistivity that are attributable to the influx of ground water in the Scholz et al. model (Figure 25) are explained by crack closure in the dilatancy-diffusionless models. Seismic activity would decrease during crack closure due to stress relaxation,[846] and resistivity would be expected to decrease whether the cracks were filled with water or closed.[845]

A different preseismic history is associated with crustal movements. The diffusion models predict maximum uplift in the epicentral region just prior to an earthquake; however, the diffusionless model developed by Stuart[845] predicts uplift (dilatancy) followed by subsidence (crack closure) before an earthquake. Consequently, geodetic measurements over epicentral regions might provide a means for testing the validity of the two models.[845]

B. Premonitory Fault Creep Model

Dieterich[852,853] has recently developed an earthquake precursor model based on *premonitory fault creep.** The fault creep parameters were derived from extensive laboratory studies in which blocks of Westerly granite (127 × 127 × 40 mm) were loaded biaxially with hydraulic rams aligned parallel to the two long dimensions of a sample. Slip occurred on a resurfaced sawcut that was oriented 45° to the two load axes. Tests were conducted at low normal stresses (40 to 180 bars). Slow- and high-speed oscillograph records revealed two stages of premonitory creep. Dieterich suggests that these creep stages prepare a pre-existing fault for seismic slip by effectively eliminating its *strength barriers* (i.e., *stress inhomogeneities*).

A Stage I creep event commences at a point on the fault and slowly propagates across most of the slip surface as a stable, long-term process. The slip rate is comparable to the external displacement rate used to load the sample. Stage II slip begins as a *velocity perturbation* when the Stage I event reaches the end of the sample. It is then driven back through the sample by elastic strain energy. As its boundary propagates along the fault, there is a decrease in frictional resistance and, hence, stress. Such a condition causes a rapid transfer of stress to the end of the Stage II zone, thus allowing the accelerated slip to propagate rapidly without additional loading. Eventually, the frictional damping of the slip boundary becomes inadequate to stabilize the process, and the boundary accelerates to near-seismic velocities. This condition culminates in seismic slip which is driven by a rapid drop in the frictional force.

It is assumed for the precursor model that (1) both stages of creep are necessary to prepare a fault for unstable slip and (2) the length of the creeping fault segment is comparable to the earthquake's source length. The second factor is a necessary requirement, for if preseismic slip zones are small and unrelated to the earthquake's source dimensions, the creep mechanism could not explain seismic precursors. In addition, it is argued, in the absence of definitive evidence, that natural faults have higher *heterogeniety levels* (i.e., larger irregularities of stress and strength) than laboratory faults, which reduces the possibility of seismic slip extending appreciably beyond the zone of preseismic creep (i.e., "runaway earthquakes").

According to the Dieterich model,[853] Stage I slip causes changes in the stress-strain field by overcoming the fault's slip-inhibiting zones where the local stress is less than the frictional strength. These field changes are thought to produce precursors such as crustal movements, electrical resistivity variations, radon emanation, b-value anomalies, and anomalous V_p/V_s. Stage II-accelerated slip, in turn, is thought to be responsible for short-term precursors such as the premonitory creep event reported by Kanamori and Cipar[854] for the May 22, 1960 Chile earthquake (M_S = 8.3). Approximately

* It will be remembered from an earlier section in this chapter that fault creep has been observed before several earthquakes and that fault creep was found to invariably precede unstable slip in laboratory experiments.

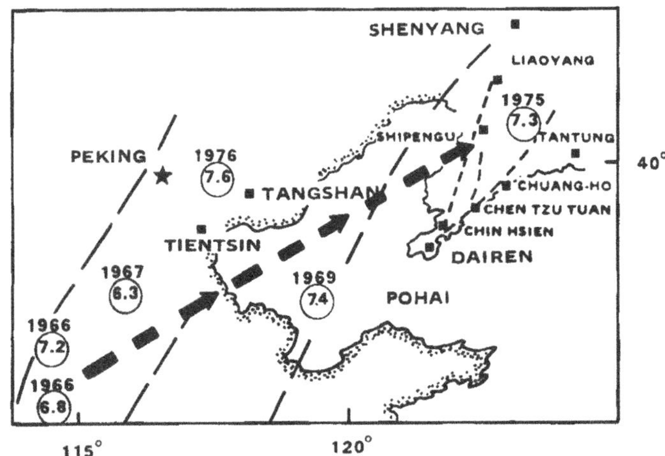

FIGURE 26. Generalized map of the northeastern region of the People's Republic of China showing faults, earthquakes with magnitudes larger than 6, and the path of a *potential moving stress force* or *deformation front*. (From Scholz, C. H., *Nature (London)*, 267, 121, 1977. With permission.)

15 min before the main shock and at about the time of a large foreshock, it appears that there was an aseismic slip episode along the potential fault plane of the magnitude 8.3 event.[854]

In reference to the V_p/V_s anomaly, Dieterich[853] proposes that (1) crack volume (and hence P-wave velocity) in a pending focal zone is more sensitive to confining pressure changes than to changes in shear stress and (2) an inhomogenous state of hydrostatic stress gives lower overall velocity than a homogenous state because the dependence of velocity on hydrostatic stress is nonlinear. Therefore, according to the model, the propagating Stage I event would progressively increase the inhomogeneity of hydrostatic stress along a fault (hence crack volume would increase), which subsequently would be responsible for progressive decreases in P-wave velocity. The recovery of V_p just before the earthquake should occur if the inhomogeneity of the hydrostatic stress was reduced. This could occur at the origination time of Stage II slip.

C. Propagating Deformation Front Model

Scholz[830] has recently proposed an hypothesis suggesting that the February 4, 1975 Haicheng earthquake and certain of its precursors could have been caused by a *moving stress force* or *deformation front* (Figure 26). The mechanics of generation for such a force is unknown, but according to Scholz,[830] its origin was probably deep seated and might have been caused by *mantle stress waves* as described by Savage[764] and Bott and Dean.[855]

The assumption that the deformation front (1) had a width of at least 300 km and (2) propagated some 1000 km through northeastern China at a velocity of approximately 110 km per year explains four types of precursors.[830] The first phenomenon was the northeast migration of large earthquakes, which commenced with two 1966 earthquakes near Hsingai (Figure 26). This migration cannot be readily explained by a change in the strain field whereby one event triggers, with some delay, the next earthquake because the distance between successive events was larger than the fault lengths associated with them, and the successive earthquakes did not occur along the same fault. Therefore, Scholz believes two possibilities remain — "either the migration was illusory or it occurred as a result of some migrating stress source."[830]

The second precursor was the increased seismic activity that commenced throughout Liaoning Province in late 1973. Scholz[830] believes that because the activity was widespread it could not have been caused by some process operative in the pending source region of the Haicheng earthquake; rather, it can be best explained by a rapid increase in the stress level throughout the entire province. Hypothetically, the stress field would have arrived in Liaoning Province in late 1973.

A third phenomenon was the successive onset of tilt anomalies at Chin Hsien, Shipengu, and Shenyang (Figure 26). Initially, the tilt vectors were north-northwest at Chin Hsien, south-southwest at Shenyang, and east at Shipengu, but the tilts rotated clockwise to the west with the passage of the deformation front. The tilt data indicate that the deformation front turned eastward about an axis south of Chin Hsien upon entering the province (Figure 26). This would explain why the tilt anomalies at Shipengu and Shenyang began after, but ended before, the anomalous activity at Chin Hsien.

The swarm sequence at Liaoyang (the fourth precursor) commenced as the maximum component of the stress force arrived at that point in late December 1974. The main earthquake occurred after the maxima had passed beyond the Haicheng region.

According to Scholz,[830] one of the major inconsistencies of the deformation front model was the July 28, 1976 Tangshan earthquake ($M_L = 7.6$) (Figure 26). Based upon the direction and propagation velocity of the front, this earthquake should have occurred in 1968. However, two conditions must be satisfied if an earthquake is to be triggered by a deformation front: (1) a zone of high stress, in which a seismic event is imminent, must lie in the path of the propagating front and (2) the stress produced by the front must be additive to the pre-existing stress. Scholz notes that the second parameter might not have been met at Tangshan.

The Tangshan earthquake occurred without an explicit prediction, although it was announced in early 1976 that a large event would either occur in the Tangshan or Peking area. However, subsequent data were "so conflicting and confusing, no final, specific prediction was ever made."[856] The earthquake was responsible for at least 650,000 deaths.

IV. EARTHQUAKE PREDICTION PROGRAMS

Government-sponsored earthquake prediction programs are now in operation in Japan, the Soviet Union, the People's Republic of China, and the U.S. These programs are very recent — the first was officially established in Japan in 1965. The U.S. program, the most recent of the four, was initiated in 1973 when federal funds specifically earmarked for earthquake prediction studies were added to the USGS's budget.[633] This section briefly describes the programs in these countries.

A. Japan

The Japanese program was proposed in 1962 when a group of seismologists informed the national government of the need for both earthquake prediction research and the development of a forecasting capability. The program was officially launched in 1965 following the destructive June 16, 1964 Niigata earthquake ($M_S = 7.5$).

The program (i.e., funding and planning) is based on a framework of 5-year increments and is now in its third 5-year segment, which commenced in 1974. For the period from 1965 through 1976, about $36 million was allocated to the program (excluding salaries). The funding was $4 million for 1975[680] but increased to $8 million for 1976.[857]

During the first part of the program, emphasis was placed on gathering base-line data related to potential long- and short-term precursors. Three centers were established to accomplish efficient data collection and processing:[745]

1. The Crustal Activity Monitoring Center of the Geographical Survey Institute is responsible for collecting geodetic and tide-gauge data.
2. The Seismicity Monitoring Center, a part of the Japan Meteorological Agency, is responsible for recording and analyzing earthquakes with magnitudes exceeding 3.0.
3. The Earthquake Prediction Observation Center, which is attached to the Earthquake Research Institute of Tokyo University, analyzes earthquakes having magnitudes less than 3.0, crustal deformation, magnetic data, and other data from university sources.

Processed data from the above centers are presented to the more recently formed Coordinating Committee for Earthquake Prediction (CCEP) for analysis. The committee is comprised of about 30 university and government specialists. Whenever there is an indication of some type of anomalous activity (e.g., crustal deformation) occurring in a particular region, it is designated as an "area of intensified observation." If, through intensified observations, the anomaly is suspected to be a precursor to a pending major earthquake, the region is ranked as an "area of concentrated observation," and efforts from various disciplines are directed at detecting a single or several short-term precursors. If discovered, a warning or prediction may be issued to the public.[745]

The Japanese prediction program was first tested (largely because of social pressures) during the Matsushiro earthquake swarm. Rikitake[182] has described this early period of the program.

> The 1965-1966 swarm earthquakes that occurred around the Matsushiro area in central Japan were certainly an epoch-making event in the history of Japanese seismology. About March 1966 when the swarm activity became so violent that local people felt an earthquake approximately every 2 minutes, a committee consisting of specialists from the Earthquake Research Institute, the Japan Meteorological Agency, the Geographical Survey Institute, and other governmental institutions was formed to investigate the Matsushiro situation and, wherever it was concluded that occurrence of a moderately large earthquake was highly probable warnings were issued to the public by the Japan Meteorological Agency. These warnings indicated the dangerous period (usually a range of a few months), a rough idea about location and possible maximum magnitude. Earthquake warnings were thus sent out to the public officially by a governmental agency for the first time in history.

Following the May 1, 1968 Tokachi-Oki earthquake, which caused considerable damage throughout northern Japan, the necessity of earthquake prediction was discussed at the cabinet level, and a second 5-year program was initiated in 1969. Unlike the first 5-year program, which was primarily geared to prediction research, this segment was oriented towards actual predictions.[745] This trend has been continued with the third increment.

By 1974, the Japanese program had 67 seismograph stations for locating earthquakes with magnitudes exceeding 3.0, 19 microearthquake observatories, 17 crustal deformation observatories (tiltmeters* and strainmeters), 12 proton-precession magnetometer stations, and one deep bore-hole seismic station. In addition, field parties can repeat first-order leveling surveys along some 20,000 km of transects every five years,[680,858] enabling anomalous uplift to be detected in any part of the country.[857]

In the bore-hole operation, the National Research Center for Disaster Prevention installed a set of high-sensitivity seismometers in a 3,500-m-deep bore-hole at Iwatsuki

* Some of the most definitive tilt anomalies have come from Japan.[680]

for microearthquake observations in the Tokyo metropolitan region. The deep borehole was necessary to suppress the "noisy" environment of the Tokyo area (e.g., industrial processes, building construction, traffic). It is possible that the system will detect small foreshocks, if any exist, before a large event takes place beneath the city. The historical record indicates that destructive earthquakes have occurred under Tokyo. Because the last one occurred in 1894, it is feared that this metropolitan region might again be hit by another of the same character.[745]

During the third 5-year program, seismic facilities have been further developed to include telemetered networks and computerized analysis facilities. In 1978, five cable-connected ocean bottom seismographs will be emplaced in the Enshu Sea, thus making it possible to precisely locate earthquakes that occur along the inner wall of the Japan trench.[680,859]

Regarding two of the more recent aspects of the program, Lapwood[857] reports that government personnel have found anomalous land deformations in the Tokai region southwest of Tokyo. Partly because of increased crustal movements since 1974, the region has been designated as an "area of intensified observation," and an intensive investigation is now centered in this region to ascertain the possibility of a large earthquake occurring there. Second, Mikumo[860] relates that it has been determined that the observation of microearthquakes (i.e., magnitudes less than 3.0) is fundamental to the understanding of seismic geography. Consequently, a nationwide project has been vigorously promoted to accomplish this task, and between 1964 and 1974, some 20,000 microearthquake epicenters were located. Their spatial arrangement clearly defines high and low frequency seismic regions which had not been previously verified by analyzing the areal pattern of larger earthquakes. These microearthquake data are also being used to search for precursory b-value changes and for temporal changes in the frequency of small shocks.[680]

In addition to the above programs, geologists and geographers are examining active faults, folding, and other crustal movements in Quaternary Period rocks to determine long-term movement rates, and laboratory studies in rock mechanics are being conducted to better understand the physical basis of earthquake prediction.[680] According to Sykes,[680] Japan will possibly become the first country to achieve routine earthquake predictions because of the following reasons.

> The Japanese have the experience gained from an 11-year national program of earthquake prediction, a vast number of trained scientists and technicians active in earthquake studies, and a relatively small geographic area to monitor, compared with the size of earthquake regions in the United States, the U.S.S.R., and China, for example. Also, most damaging earthquakes in Japan occur at shallow depths within the islands. These shocks generally tend to be more damaging even though they are smaller than the great earthquakes located off the east coast of Japan. Shallow earthquakes within lithospheric plates, such as those within the Japanese islands, are the sources of many of the precursory effects detected thus far. The shallow nature of the sources and the fact that the source regions can be readily surrounded by instruments, make it much easier to monitor possible precursory changes than if the earthquakes were located off the coast.*

Details of the Japanese prediction program are described in detail in Rikitake's recently published book entitled *Earthquake Prediction*.[861]

B. Soviet Union

Soviet scientists were the first to conduct earthquake prediction research. For example, the initial step to study precursors was initiated in 1938 by V. F. Bronchkovsky, who installed a series of tiltmeters in the Alma Ata region.[862] Following the destructive

* Epicenter plots (1962 through 1969) for Japan are depicted in Figure 83 in Volume I, Chapter 2.

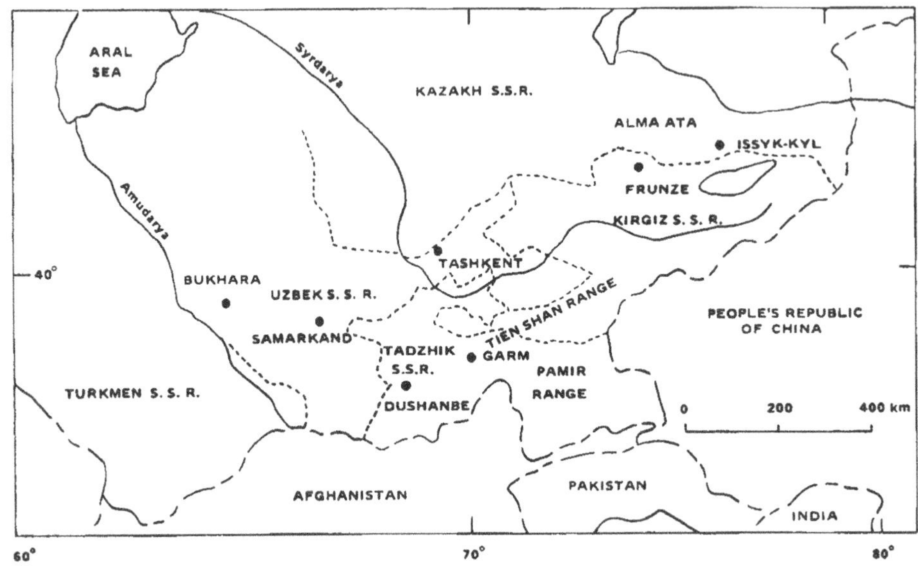

FIGURE 27. Central Asian seismic zone of the Soviet Union. (From Wesson, C. V. K. and Wesson, R. L., *Earthquake Inf. Bull.*, 7, 10, 1975.)

1949 Khait earthquake (M_L = 7.5) in the Garm region, the Complex Seismological Expedition was formed "to record and study the spatial distribution of earthquakes, their variations with time, and other statistical properties."[680] According to Sykes,[680] the research "in the Garm region was later expanded to include searches for precursory phenomena associated with earthquakes." Governmental prediction and hazards reduction programs were officially launched after the April 26, 1966 Tashkent earthquake. Although the earthquake was only of moderate magnitude (M_L = 5.2), it occurred directly beneath the city of Tashkent.[680] This event destroyed or damaged 85,000 dwellings and was responsible for 1800 deaths.[863]

Earthquake precursor studies are concentrated in the country's two most active seismic regions: Central or Middle Asia and the Kamchatka Peninsula. The Central Asian region, the most densely populated seismic zone, consists of the Kazakk, Uzbek, Tadzhek, and Kirgiz republics (Figure 27). The Alma Ata, Garm, and Tashkent accounts are from this part of the Soviet Union (Figure 27). Its tectonic environment is characterized by a north-south compression believed to be associated with the continental collision of the Indian subcontinent and Asia.[680] Epicenter plots (1962 through 1969) for this region are depicted in Figure 82 in Volume I, Chapter 2.

The Kamchatka region is in the western Pacific Basin. Its seismic regime is similar to that of the Aleutians and Japan, i.e., rapid underthrusting of island arc structures by the Pacific plate.[680] Epicenter plots (1962 through 1969) for this region of the Soviet Union are depicted in Figures 77 and 83 in Volume I, Chapter 2.

Precursor investigations in Central Asia are concerned with seismic wave velocities, foreshock activity,electrical resistivity, geomagnetism, geodetic surveys, crustal tilt, and geochemical anomalies. According to Savarensky.[742] these types of investigations have been intensified in the active areas of the Pamir and Tien Shan ranges (Figure 27) where future earthquakes would threaten the capitals of the Central Asian republics. Field studies are being carried out by the Institute of Physics of the Earth, Academy of Sciences of the U.S.S.R. and the Academies of Sciences of the Kazakh, Kirghiz, Uzbeck, Tajik, and Turkmen republics.[742]

Attempts are being made in Kamchatka to use anomalies in the electrotelluric field, seismic wave velocities, and the seismic cycle established by Fedotov[636] for predicting earthquakes. The cycle is determined by the frequency and magnitude of small events within the time interval between two strong ($M_L \geq 7$) earthquakes.[742] Other types of precursor studies are restricted because the epicentral regions for many of the earthquakes are on the ocean floor. Investigations are being conducted by scientists from the Institute of Volcanology, Petropavlovsk-Kamchatsky and the Institute of Physics of the Earth.

Based upon a recent visit to the Soviet Union, Dieterich and Brace[864] reported that most of the laboratory research pertaining to earthquake mechanics and prediction is being carried out at the Institute of Physics of the Earth in Moscow. The institute houses five laboratories, under the directorship of V. I. Myachkin, where these studies are concentrated. The five laboratories specialize in "theory, modeling, mechanical and electrical processes, high-pressure studies, and studies of the state of stress and fracture phenomena."[864]

Although no specifics have been made available concerning the financial scale of the Soviet prediction effort, Rikitake[182] reported in the late 1960s that about 80 and 15% of the yearly earthquake prediction budget were spent for the Central Asia and Kamchatka projects, respectively. The remaining 5% was devoted to laboratory research. Detailed descriptions and results of the investigations for Central Asia have been prepared by Sadovsky et al.,[678] Savarensky,[862] and Sadovsky and Nersevov[865] and for Kamchatka by Fedotov et al.[710,866,867]

C. People's Republic of China

To date, the most sophisticated and successful earthquake prediction program is in the People's Republic of China, a country that has lost more than two million of its citizens to earthquake disasters (Table 1 in Volume I, Chapter 1). The high casualty rate is due, in large measure, to the country's almost total lack of earthquake-resistant housing. Unreinforced brick construction is common in urban areas, and mud and masonry construction with heavy tile or mud roofs is common in rural areas.[868] The structural design of rural housing has changed little since the 16th century, and more than 600 million people now live in such housing.[680]

Economic constraints make it impossible for the Chinese to replace the approximately 30% of the country's total housing that occupies earthquake-prone regions.* Hence, the government has adopted a policy to develop reliable earthquake prediction techniques that will make it possible for people to be evacuated from their homes before destructive earthquakes strike.[869] Following a successful prediction of a destructive event, reconstruction would occur in the affected region without being a burden to the national economy.[680]

The impetus to establish a national program in earthquake prediction came as a result of the March 8 and 22, 1966 Hsingai, Hopeh Province, M_L = 6.8 and 7.2 earthquakes (Figure 26). Prime Minister Chou En-lai visited the striken area and declared that prediction research was to be given the highest priority.[869] The Hsingai earthquakes occurred at about the beginning of the Great Proletarian Cultural Revolution (1966 to 1969), and the ensuing prediction program was strongly influenced by political philosophy. For example, criticism was leveled at the "elitist" scientific community for having been unconcerned with serving the needs of the people. Therefore, the prediction program was intended "to demonstrate how scientists can combine with workers, peasants, and soldiers to solve a problem of great national concern,"[825] or as Shapley[870] notes, the program reflects the Maoist ideal of a "people's science."

* Epicenter plots (1962 to 1969) for the People's Republic are depicted in Figure 82 in Volume I, Chapter 2.

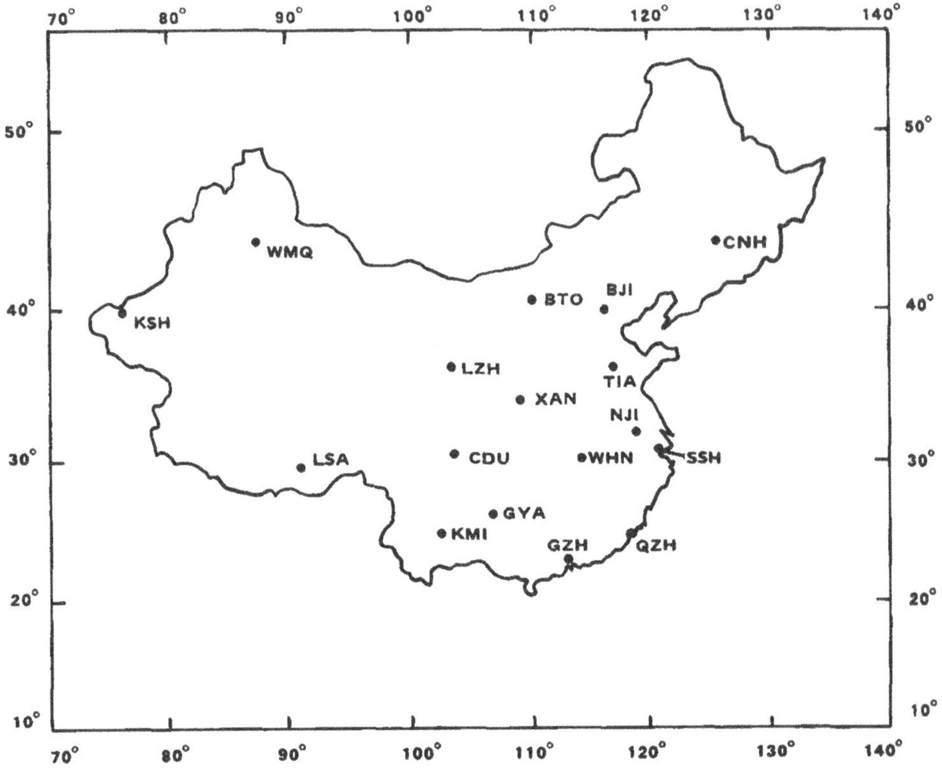

FIGURE 28. Seventeen standard seismograph (three component short- and long-period instruments) stations in the People's Republic of China. Station codes include: KSH = Kuhsi, WMQ = Urumchi, LSA = Lhasa, KMI = Kunming, GZH = Canton, QZH = Ch'uan-chou, GYA = Kweiyang, CDU = Chengtu, WHN = Wuhan, SSH = Shanghai, NJI = Nanking, XAN = Sian, LZH = Lanchow, TIA = Taian, BJI = Peking, BTO = Paotow, and CNH = Changchum. (From Hamilton, R. M., *Earthquake Inf. Bull.*, 7, 5, 1975.)

Consequently, the prediction program is comprised of about 10,000 professionals (including administrative and technical support personnel) and more than 100,000 part-time amateur observers.[825] The amateur observers are usually from some civil unit (e.g., schools, factories, communes) and are primarily responsible for monitoring water well levels, electrical resistivity, ground tilt, and animal behavior. These amateur groups also are involved in a mass education program to foster an understanding of earthquake prediction and protection measures.

Various types of instrumentation are deployed throughout the country. For example, the Institute of Geophysics of the Chinese Academy of Sciences (Academia Sinica), with headquarters in Peking, is responsible for the operation of 17 standard seismograph observatories (Figure 28), 250 regional and local seismograph observatories (including an 8-station network around Peking), and 8 geomagnetic observatories. There are at least 5000 *observation points* scattered throughout the earthquake prone regions. A point might consist of an observatory or a single well for monitoring radon emanation.[871]

Most Chinese scientists believe that earthquakes are preceded by a wide range of precursory phenomena and that they occur in a variety of tectonic environments. As noted by Sykes and Raleigh,[675] the following precursors that have been reported before large earthquakes are being studied in extensive field programs: (1) crustal deformation, (2) seismic wave velocities, (3) frequency of occurrence of small earthquakes, (4) radon flux from well water and springs, (5) water level and temperature in wells, (6) magnetic field strength, (7) natural telluric current, (8) active electrical resistivity, and (9) abnormal animal activity. Precursors are usually categorized as long term (e.g., years), intermediate term (e.g., months to weeks), and short term (e.g., days to minutes). To date, little effort has been directed toward theoretical studies (e.g., dilatancy) that could provide plausible explanations for the occurrence of precursors. Scientists uphold the assertion of Chairman Mao-Tse-tung that "purely theoretical exercises cannot advance useful scientific knowledge."[675]

The Institute of Geophysics carries out most of the seismological research. The institute was reorganized during and following the Cultural Revolution with earthquake prediction established as the highest priority goal. Consequently, the program of the institute is organized into four seismology and general geophysics sections and one geomagnetism section. The seismology and general geophysics sections have the following responsibilities: (1) determination of regional seismicity characteristics (2) seismic risk analysis, including the locales for short-range prediction studies; (3) investigations into the structure of the earth's crust and interior; and (4) improvements to seismic instruments and observation techniques. The geomagnetism section is currently conducting a magnetic survey of the entire country and investigating geomagnetic methods for earthquake prediction.[872]

Data for the institute's programs are gathered in the field by provincial seismological brigades. A single brigade has its own observatory and serves as a regional center for accumulating data from a province or group of provinces.[869] Brigade personnel also provide technical training and advice to amateur observation groups.[825]

Seismogeologic tasks are also being carried out at the Institute of Geology and the electrical properties of rocks are being investigated at Peking University.[872] Earthquake-animal behavior research is being conducted at the Institute of Biophysics in Peking and at Peking University.[825]

Virtually nothing was known about the Chinese program until quite recently. The constraints of the Cultural Revolution prohibited the publishing of scientific journals from 1966 until 1973, and for most of this same period, western scientists were barred entry. Wilson,[108] a Canadian geophysicist, spent 3½ weeks in the country as a guest of the Academy of Sciences and was one of the first westerners to report on the massive efforts underway to predict earthquakes. Additional information has been made available to recently visiting American scientists. Bolt[758] visited the country in 1973; under sponsorship by the Committee on Scholarly Communication with the People's Republic of China (CSCPRC),* two American delegations of government and university scientists visited the country in 1974[873] and 1976.[825]

The first American delegation was informed by Chinese authorities that 11 earthquakes had been successfully predicted by 1974. Each prediction resulted in the evacuation of people from their homes.[868] Based upon information supplied to the second delegation, the successful predictions of the February 4, 1975 Haicheng earthquake (M_L = 7.3) and the May 29, 1976 earthquake (M_L = 6.9) that struck near the communities of Tengtsung and Lungling in the western part of Yunnan Province[825] can be added to the 1974 total. The authorities also readily admit, without offering details, to a number of prediction failures. Undoubtedly, the largest event to have escaped a successful prediction was the July 28, 1976 Tangshan earthquake (M_L = 7.6) which was responsible for at least 650,000 deaths.

* An endeavor of the National Academy of Sciences, the Social Science Research Council, and the American Council of Learned Societies.

Although figures have not been made available, Sykes[680] believes that the yearly cost of the Chinese prediction program could be in the $50 to $100 million range. Extensive descriptions of the earthquake prediction effort in the People's Republic are found in the reports of Bolt,[758] the 1974 American Seismology Delegation,[873] and the 1976 Haicheng Earthquake Study Delegation.[825] The organizational structure for issuing warnings to the public is described in Volume III, Chapter 2.

D. United States

It is not known when scientific prediction studies actually began in the U.S.; however, the March 27, 1964 Alaskan earthquake (M_s = 8.5) stirred considerable interest in a prediction capability, a view fortified by the February 9, 1971 San Fernando earthquake (M_s = 6.5). Prior to the former event, most American scientists either ignored the prediction topic or were reluctant to publish their findings. Oliver[874] maintains that, in part, the latter could be due to the nationwide publicity given to prediction by seers and mystics based on no scientific data. Unfortunately, predictions based on quackery are still being made in California.

The first concerted effort for organizing a prediction program at the federal level occurred in 1965 when, at the request of President Lyndon B. Johnson, the Ad Hoc Presidential Panel on Earthquake Prediction was formed.[875] The panel of eminent scientists proposed a major 10-year program on earthquake prediction that called for: (1) the installation of microseismicity arrays; (2) the development and installation of tiltmeters, strainmeters, magnetometers, gravimeters, and surveying devices; and (3) the installation of telemetry networks. Emphasis was to be placed on fault studies in California and Alaska, and the 10-year cost of the program was estimated at $137 million.[875]

The USGS and the U.S. Coast and Geodetic Survey began small-scale prediction studies the following year, as did a number of university scientists with funds largely provided by the National Science Foundation.[633] Additionally, the Secretary of the Interior established an advisory panel of 10 to 15 nongovernment scientists to counsel the USGS "on the feasibility, appropriateness, and scientific value of its earthquake program so that it best serves the national interest."[876] This panel continues to meet twice a year.

In 1973, federal funds specifically earmarked for prediction research were added to the Geological Survey's budget and the Office of Earthquake Studies (OES) was created within the Geologic Division of the USGS to better serve the federal earthquake research effort. Within OES, scientists attached to the Branch of Earthquake Mechanics and Prediction, with headquarters in Menlo Park, California, conduct most of the Geological Survey's field, laboratory, and theoretical prediction research. The other OES branches are the Branch of Seismic Engineering (San Francisco), the Branch of Earthquake Hazards (Golden, Colorado), and the Branch of Seismicity and Earth Structure (Menlo Park). Many of the prediction procedures follow guidelines proposed by the 1966 Ad Hoc Presidential Panel on Earthquake Prediction.

To date, field research is primarily centered along the central San Andreas fault zone, one of the most seismically active sections. There, the USGS operates a prototype network of various instruments for the simultaneous detection of possible earthquake precursors, including seismographs (Figure 39 in Volume I, Chapter 2), tiltmeters (Figure 9), magnetometers, and dipole-dipole variometers; geodimeter lines are normally surveyed once a year. A similar network has been recently installed in the region of the Palmdale bulge in southern California (Figure 11) to search for potential precursors.

Following the passage of the *Disaster Relief Act of 1974*, the responsibility for warning of geologic catastrophes was delegated to the Director of the USGS. As a response to this act, the USGS has established the *Earthquake Prediction Panel* "to be responsible for reviewing data that could warn of an earthquake and for recommending that a prediction be issued,"[877] with the Director issuing the actual prediction. The *California Earthquake Prediction Evaluation Council* is an organization with similar responsibilities at the state level.[878] These groups are described in Volume III, Chapter 2.

Federal funds allocated for in-house and grant programs to universities, private industry, and state geological surveys totaled $10,637,000 in Fiscal Year 1976: Geological Survey — $5,000,000; National Science Foundation — $4,252,000; National Aeronautics and Space Administration — $1,300,000; and the Nuclear Regulatory Commission — $85,000.[633]

Because of the enactment of the *Earthquake Hazards Reduction Act of 1977*, federal funding for earthquake prediction and other types of mitigation research will drastically increase in the period fiscal years 1978—1980. During this period, more than $200 million will be committed for accelerated investigations. The funding will be shared by the USGS and the National Science Foundation.[879] The above law is described in detail in Volume III, Chapter 3.

The U.S. has been participating in joint programs with Japan and the Soviet Union. Scientists from the U.S. and Japan have met five times since 1964 in an "exchange of information" program concerned with the present status and future prospects of the prediction efforts in each country. The multi-day seminars are a part of the U.S.-Japan Cooperative Science Program and are sponsored by the National Science Foundation and the Japan Society for the Promotion of Science.[633,859,880]

As a result of the Agreement on Cooperation in the Field of Environmental Protection of 1972, the U.S.-U.S.S.R. Working Groups on Earthquake Prediction have established joint field, laboratory, and theoretical studies on prediction and related fields;[633,881] the agreement was renewed in 1977 for an additional five years.[882] The program is co-chaired by M. A. Sadovsky, Institute of Physics of the Earth, and R. E. Wallace, USGS.[881]

The Panel on Earthquake Prediction,[633] (National Academy of Sciences), describes the organization and results of the program:

> The work has been organized into four areas: (1) field investigations of earthquake prediction; (2) laboratory and theoretical investigations of the earthquake source; (3) mathematical and computational prediction of places where large earthquakes occur and evaluation of seismic risk; and (4) engineering-seismological investigations. Successful projects to date include the establishment of a joint seismograph network near Garm, places where large earthquakes occur and evaluation of seismic risk; and (4) engineering-seismological investigations. Successful projects to date include he establishment of a joint seismograph network near Garm, Tadzhik SSR, to search for seismic forerunners to large earthquakes; the establishment of a joint seismograph network around the Nurek Reservoir, Tadzhik SSR, to study reservoir-induced seismicity; the study of the spectral content of earthquakes using the Soviet "frequency-selecting" seismograph system and American broad-band, digitally recording equipment; collaborative laboratory studies of the effects precursory to failure and sliding in rock samples; the application of pattern-recognition techniques to the prediction of earthquakes; the establishment of a joint strong-motion network in the Tadzhik SSR; and the collaborative instrumentation and testing of buildings subjected to earthquake-like motions simulated by explosions.

V. FUTURE PROSPECTS

Upon reviewing the research efforts in predicting earthquakes in Japan, the Soviet Union, the People's Republic of China, and the U.S., the members of the Panel on Earthquake Prediction[633] summarized the status of prediction efforts as of 1976.

1. Earthquake prediction holds great potential for saving lives, reducing property damage, enhancing the safety of critical facilities, and helping make possible more-rapid restoration of normal living after an earthquake.
2. Anomalous physical pehnomena precursory to some earthquakes have been clearly identified.
3. The physical nature of precursory phenomena is complex, and current models to explain them are crude; improvements of these models will require considerable effort in the field and laboratory, as well as in theoretical studies.
4. Some small earthquakes have been predicted in a scientifically credible way, and most researchers are optimistic that we will eventually be successful in predicting larger earthquakes.
5. Of about ten types of recognizable phenomena thought to be precursory to earthquakes, some may, in fact, be due to other causes and yield false alarms. Successful routine prediction will probably require the use of several techniques.
6. At present, the ability to detect and locate an impending earthquake requires a dense distribution of instruments in the quake area. Improved observational networks in areas of high earthquake probability are mandatory if we are to gain the fundamental knowledge on which to build an effective earthquake-prediction program.
7. Predictions of earthquakes should specify time, magnitude, place, and probability. However, even a statement that does not specify time or magnitude, or a statement that an earthquake will not occur in a particular place or at a particular time, would be beneficial.
8. Neither the present state-of-the-art nor the present distribution of instrumentation permits socially useful predictions on a routine basis. Therefore, at this time, an expression such as "area of intensive study," as used in Japan, might reflect more accurately the confidence level of interpretations of the observed phenomena in some areas than would an actual prediction.
9. A scientific prediction will probably be made within the next five years for an earthquake of magnitude 5 or greater in California. With appropriate commitment, the routine announcement of reliable predictions may be possible within 10 years in well instrumented areas, although a large earthquake may present a particularly difficult problem. The apparent public impression that routine prediction of earthquakes is imminent is not warranted by the present level of scientific understanding.
10. Until formal procedures for issuing predictions have been established, predictions made by responsible scientists should be accompanied by sufficient backup data for full evaluation by the scientific community.
11. During the development of an earthquake-prediction-and-warning capability, there will be unavoidable errors and false alarms. The public must be made aware of this prospect, and the development of any procedure to issue warnings must accommodate it. Even the ultimate system probably will not be infallible.
12. The rate of development of a reliable earthquake-prediction capability operating on a routine basis will depend to a large extent on the amount, rate, and deployment of funding. Progress in improving the state-of-the-art in the early growth period will be particularly sensitive to the level of support. The Panel believes that an effective program will require a 10-year commitment of effort, and that a large increase to several times the current annual Federal expenditures would be cost effective and would be in the national interest.
13. The scientific and technical aspects of earthquake prediction have advanced to the point at which the development of systems for associated societal response should be addressed promptly in a formal manner. A prediction capability will be of little value if societal response procedures are not formulated concurrently.
14. In a realistic attack on the earthquake-hazard problem, the development of an earthquake-prediction program and the upgrading of earthquake-engineering design and construction are complementary and equally necessary, and should be carried on at the same time.

Chapter 2

EARTHQUAKE CONTROL

I. INTRODUCTION

In addition to the possiblity of being able to *predict* earthquakes on a reliable basis in the future, it might be possible to *control* the time of occurrence and size of shallow focus earthquakes along fault zones. Beginning in the 1960s, it was accidentally discovered that small- to moderate-sized earthquakes could be triggered artificially by (1) *the detonation of underground nuclear devices* and (2) *the underground injection of fluids into stressed rocks at high pressure.** Both mechanisms have been looked upon as effective methods for reducing earthquake hazards by the premature release of accumulating strain energy via numerous small crustal displacements along existing faults. Ideally, this should inhibit the accumulation of sufficient energy to produce infrequent, but large, ruinous earthquakes.

II. UNDERGROUND NUCLEAR EXPLOSIONS

It has been established that large underground nuclear explosions detonated primarily at the Nevada Test Site** are followed by a series of shallow-focus earthquakes that occur: (1) in the vicinity of the explosion cavity (i.e., cavity collapse earthquakes) and (2) along pre-existing faults, usually within 10 km of ground zero.[90,883-892] The displacements and associated earthquakes along faults result from an explosion (transient strain) serving as a mechanism for triggering the release of natural tectonic strain in decreasing amounts from the shot point.[893-906]

Edward Teller, a prime architect of nuclear science, was one of the first scientists to speculate in general terms on the possibility of using underground nuclear explosions to control earthquakes. More specifically, in a 1969 plan proposed by Emiliani et al.,[883] a series of 1- to 10-Mton nuclear devices would be placed in 3000- to 5000-m deep wells spaced at 20- to 50-km intervals along an active fault zone. With sequential detonation, the accumulated strain would be released, and the procedure would be repeated at 10- to 25-year intervals. The western region of the Aleutian Islands was proposed as a potential area to test this plan.

However, for several reasons, the nuclear detonation mechanism for earthquake control has been abandoned in the last few years. One problem centers on the huge expenses that would be involved in drilling initial and successive bore holes (new bore holes would be needed after each blast) to sufficient depths to reach the active hypocentral zones of many shallow earthquakes.[90] Second, due to uncertainties in determining stress levels in the crust, a nuclear detonation might trigger a much larger earthquake than anticipated. A third problem is public concern regarding any aspect of nuclear explosions.

III. FLUID INJECTION

The notion that earthquakes could be controlled by underground fluid injection at

* Induced seismicity also can be caused by rock failures in mines and apparently by the impoundment of large quantities of water in reservoirs. The latter topic is discussed in Volume III, Chapter 1.

** The first U.S. underground nuclear device (1.7 kton) was fired at the Nevada Test Site on September 19, 1957.[90]

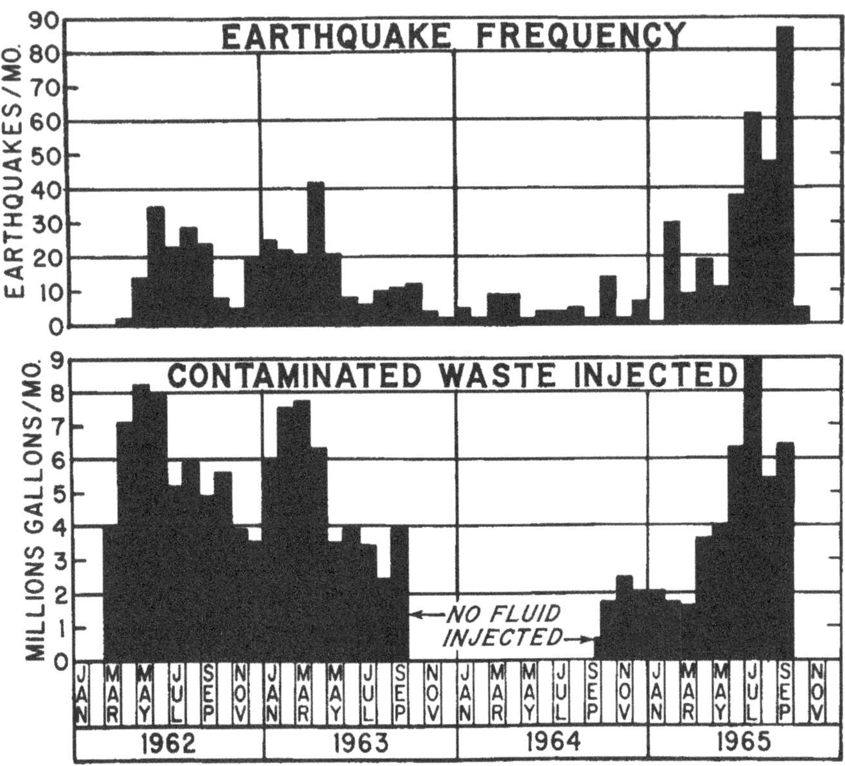

FIGURE 1. Earthquake frequency compared to fluid injection rates at the U.S. Army's Rocky Mountain Arsenal well, Colorado. (From U.S. Geological Survey, *Earthquake Inf. Bull.*, 5, 6, 1973.)

high pressure began in 1966 when David Evans,[907] a consulting geologist, reported that the injection of liquid-chemical wastes in a disposal well at the U.S. Army's Rocky Mountain Arsenal, Colorado had triggered hundreds of small earthquakes in the immediate area. Because of the closeness of the arsenal to the Denver metropolitan area, these seismic events have been called the *Denver earthquakes.* Planned earthquake control experiments have been successfully completed at the Rangely oil field, Colorado and at Matsushiro, Japan. In addition to the Denver earthquakes, the evidence suggests that seismic events have been inadvertently triggered by fluid injection at Dale, New York and Los Angeles, California.

A. Denver Earthquakes

Since it was no longer possible to use evaporative ponds, a 3671-m deep well was drilled into the highly fractured Precambrian gneiss in the Denver Basin by the U.S. Army in 1961 to dispose of contaminated wastewater produced in the manufacture of nerve gas at the Rocky Mountain Arsenal. Pressure injection was started in March 1962, and between April 1962 and October 1965, 710 earthquakes occurred within about 11 km of the disposal well; local magnitudes (M_L) ranged from 0.7 to 4.3. This was a marked difference from historic activity for, as Ives[908] noted, only 11 earthquakes had been felt in the Denver area between 1840 and 1962.

Using Army records of times and amounts of fluid injected, Evans[907] correlated earthquake frequencies with waste injection amounts and pressures (Figure 1). There were five characteristic periods of the waste injection program:

1. April 1962 to April 1963, high injection at medium pressure
2. May 1963 to September 1963, medium injection at medium pressure
3. October 1963 to September 1964, no injection
4. September 1964 to March 1965, low injection at gravity pressure
5. April 1965 to September 1965, high injection at high pressure

Evans[907] made the following observations:

1. From March 1962 to the end of September 1963, the injection program was often shut down, and a monthly correlation did not appear to be significant.
2. High injection months of April, May, and June 1962 seemed to correlate with the high earthquake frequency months of June, July, and August.
3. High injection months of February and March 1963 may correlate with the high earthquake frequency month of April.
4. The no injection period from September 1963 to September 1964 coincided with a period characterized by a low earthquake frequency.
5. The period of low volume injection by gravity flow (September 1964 to April 1965) was characterized by 2 months (October and February) of greater earthquake frequency than that experienced the preceding year.
6. The strongest correlation was during the months of June through September 1965, a period characterized by the pumping of 1350 ℓ of contaminated water for 16 to 24 hr a day at pressures of from 800 to 1050 psi.

Evans[907] further suggested that the increased fluid pressure within the reservoir had triggered the earthquakes in a manner proposed earlier by Hubbert and Rubey,[909] geologists who explained the role of fluid pressure in lowering the normal stress across the failure plane of an overthrust fault. They showed that the shearing stress needed to cause slippage could be made very small by simply increasing fluid pressure which would reduce the coefficient of friction. Therefore, if the fluid pressure was sufficiently high, large rock masses could move over a nearly horizontal surface or blocks under their own weight could slide down more gentle slopes than otherwise would be possible.[910]

The basic concept of the *Hubbert-Rubey mechanism* can be demonstrated by the following relationships.[911] The shearing stress of rocks is expressible by

$$\tau = \tau_o + \mu\sigma_n \tag{1}$$

where τ = failure strength, τ_o = intrinsic strength, μ = coefficient of friction, and σ_n = normal stress across the plane of failure. An increase in fluid pressure (P) reduces the effective normal stress, thereby reducing the strength of a fault zone by

$$\tau = \tau_o + \mu(\sigma_n - P) \tag{2}$$

Raleigh[910] has developed a simple laboratory procedure for dramatically illustrating the Hubbert-Rubey mechanism.

Evans[907] proposed that when the fluid pressure in the gneiss fractures was increased by the Army's injection program, the resistance to sliding along fracture planes in the basement rocks was reduced, and block movements occurred, resulting in the Denver earthquakes. He concluded:

. . . it is interesting to speculate that the principle of increasing fluid pressure to release elastic wave energy might have an application in the subject of earthquake modification. That is, it might someday be possible to relieve the stresses along some fault zones in urban areas by increasing the fluid pressures along the zone using a series of injection wells. The accumulated stress might thus be released at will in a series of non-damaging earthquakes instead of eventually resulting in one large event that might cause a major disaster.

Healy et al.[912] were a part of an investigative team assembled by the U.S. Army Corps of Engineers to evaluate the Evans theory of fluid injection and earthquakes at the Rocky Mountain Arsenal. Upon analyzing the seismic data for the Denver area between 1962 and early 1967, the group found general agreement with the Evans theory. They noted that:

1. No seismic similarity was found for the period prior to 1962 and that which followed the termination of the fluid injection program.
2. Epicenters were located in a fairly narrow zone about 10 km long, with the disposal well situated roughly in the center; focal depths were just below the bottom of the injection well.
3. The probability of finding an earthquake swarm in a randomly selected 65-km^2 area (approximate size of the epicentral region) would be 1 in 4150.
4. The probability that an earthquake sequence would commence within 7 weeks of the beginning of fluid injection in the disposal well was 1 in 600.
5. The joint probability that the earthquakes would be so closely associated in time and space with the disposal well was 1 in 2,500,000. Therefore, the occurrence of a natural sequence of earthquakes so closely related to the disposal well would be an extremely unlikely coincidence.

Due to the possible connection between fluid injection and the triggering of earthquakes and the closeness of the Denver metropolitan area to the arsenal, the injection program was permanently stopped on February 20, 1966. Although a few earthquakes occurred a year or two later, the seismic activity in the Denver area has essentially ceased.[913]

B. Rangely Oil Field, Colorado Earthquakes

As noted in a 1973 U.S. Geological Survey (USGS) publication:[913]

Earth scientists, hearing about the Denver quakes and their cause, were attracted by the possibility that earthquakes might be controlled. If they could be turned on, perhaps they could be turned off, or perhaps their magnitudes could be kept to safe levels. Obviously, a well-planned experiment was required, but the expense of such an experiment would be staggering.

Fortunately, it was discovered that it might be possible for the USGS to conduct an inexpensive field experiment in earthquake control at the *Rangely oil field* in northwestern Colorado (Figure 2). At Rangely water under pressure was being injected into converted oil wells to force petroleum from the Weber sandstone towards lower pressure wells, where it was then pumped to the surface (i.e., *waterflooding method* for the *secondary recovery* of petroleum). Oil wells near the periphery of the field were the first to be converted to water-injection wells. Some 97 wells had been converted for injection purposes by September 1965; the number had increased to 202 by September 1969.[914]

In 1967, USGS scientists learned that a large number of small earthquakes were occurring in the vicinity of the Rangely oil field. These seismic data had been recorded by an U.S. Air Force seismic array located 60 km east of Rangely.[913] Consequently, the USGS installed four portable seismographs around the periphery of a portion of the field, and 20 small earthquakes were recorded in an 8-day period in November

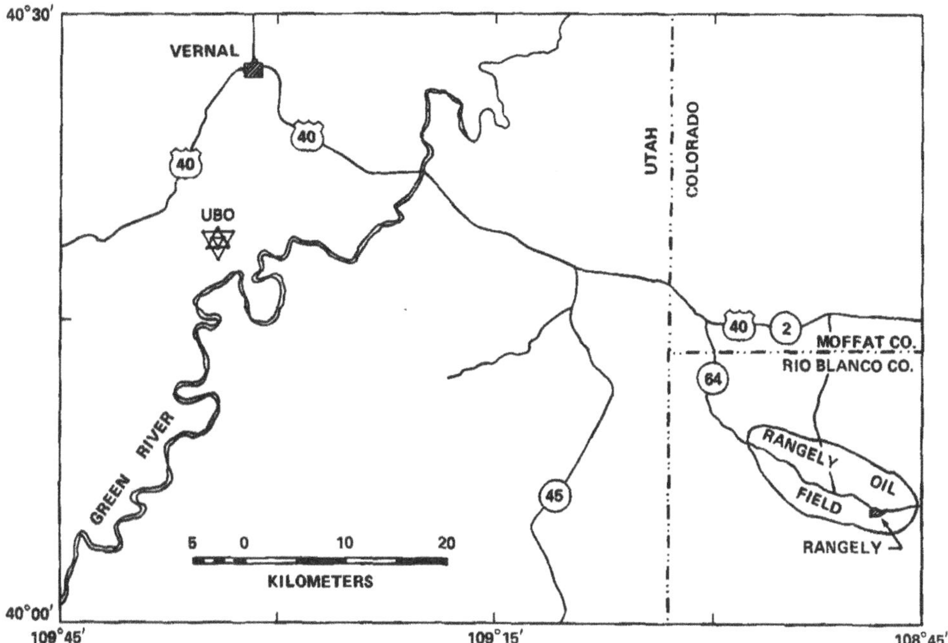

FIGURE 2. Location of the Rangely oil field, Colorado and the Uinta Basin Seismological Observatory, located approximately 65 km west northwest of the field. (From Gibbs, J. F., Healy, J. H., Raleigh, C. B., and Coakley, J., *Bull. Seismol. Soc. Am.*, 63, 1558, 1973. With permission.)

1967 (Figure 3). The epicenter plots revealed a pattern of activity that was coincidental to two areas of the field where fluid pressures accountable to waterflooding were the highest.[914]

These preliminary findings suggested that the earthquakes were related to the repressurization of the reservoir strata. A cooperative agreement was reached between the Chevron Oil Company (operator of the field) and the USGS and supported by the Advanced Research Projects Agency of the Department of Defense to determine if earthquakes could, in effect, be turned on and off at will by varying the field's fluid pressure.[913,915] The full-scale experiment for a restricted part of the Rangely oil field began in September 1969 and, according to Raleigh et al.,[915] was planned as follows:

1. Subsequent to recording seismic activity from 14 seismographs for 1 year, the fluid pressure in the vicinity of the earthquakes would be reduced by *backflooding* (i.e., withdrawing) water from four injection wells.
2. If the fluid pressure reduction caused a diminution in seismic activity, the pressure would be increased again by injection, and the cycle repeated.
3. Measurements of reservoir pressures in nearby wells would be used to estimate the spatial disribution of pressure concurrent with the cycles of injection and withdrawal. Field and laboratory measurements of *in situ* stress and the fictional properties of the reservoir rock would be used to test the Hubbert-Rubey mechanism by comparing these observations with the predicted fluid pressure level needed to trigger earthquakes on pre-existing fractures.

While the above investigation was in its initial stages, Gibbs et al.[914] of the USGS provided additional evidence suggesting that earthquakes were being triggered artificially at the Rangely oil field. They conducted a seismic frequency-injection study for

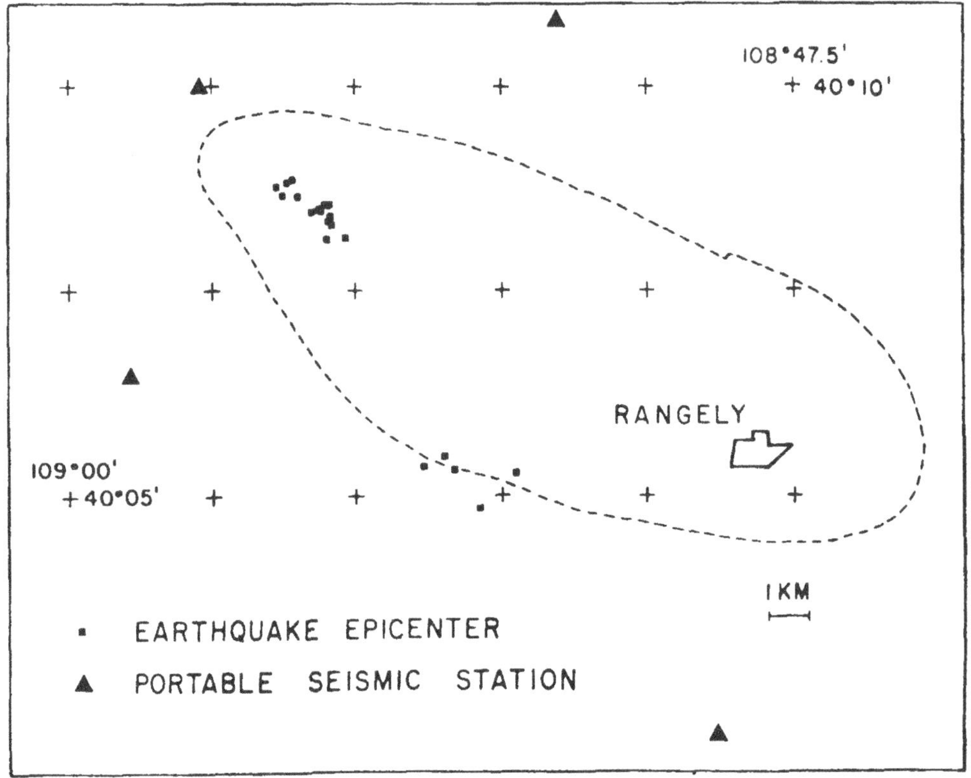

FIGURE 3. Location of portable recording seismographs and earthquake epicenters in the vicinity of the Rangely oil field, Colorado for an 8-day recording period in November 1967. (From Gibbs, J. F., Healy, J. H., Raleigh, C. B., and Coakley, J., *Bull. Seismol. Soc. Am.*, 63, 1560, 1973. With permission.)

the entire field by using seismic data recorded during a 7-year period at the Uinta Basin Observatory, located 65 km west-northwest of the field (Figure 2). Between November 1962 and January 1970, it was determined that 976 earthquakes had occurred in the field and nearby areas, of which 320 had local magnitudes exceeding 1. Upon comparing the yearly number of earthquakes with the fluid injected per year, Gibbs et al.[914] found that the absolute value of water injected per year did not appear to affect earthquake frequency; rather, changes in the quantity of water injected appeared to correlate with changes in earthquake frequency. If injection increased, seismic activity also increased (Figure 4). This relation existed for all years except 1969 (Figure 4).

The following discussion is concerned with the results of the USGS-Chevron cooperative program in earthquake control previously described. From October 1969 to November 1970, water was injected into the four test wells, and the bottom-hole pressure was raised from 235 to 275 bars.* During this period, more than 900 earthquakes ($M_L \geqslant -0.5$) occurred in the field. Epicenter plots indicated that most of the events were confined to two areas: one in the immediate vicinity of the four injection wells and the second immediately to the southwest (Figure 5). Some 367 events occurred within 1 km of the bottom of the injection wells (Figure 6). Hypocenters averaged 3.5 km for the southwest cluster and approximately 2.0 to 2.5 km beneath the well heads or within the injected horizon. The events comprising both clusters were located about the vertical zone of a strike-slip subsurface fault along those segments where fluid

* One bar equals atmospheric pressure at sea level or approximately 14.7 psi.

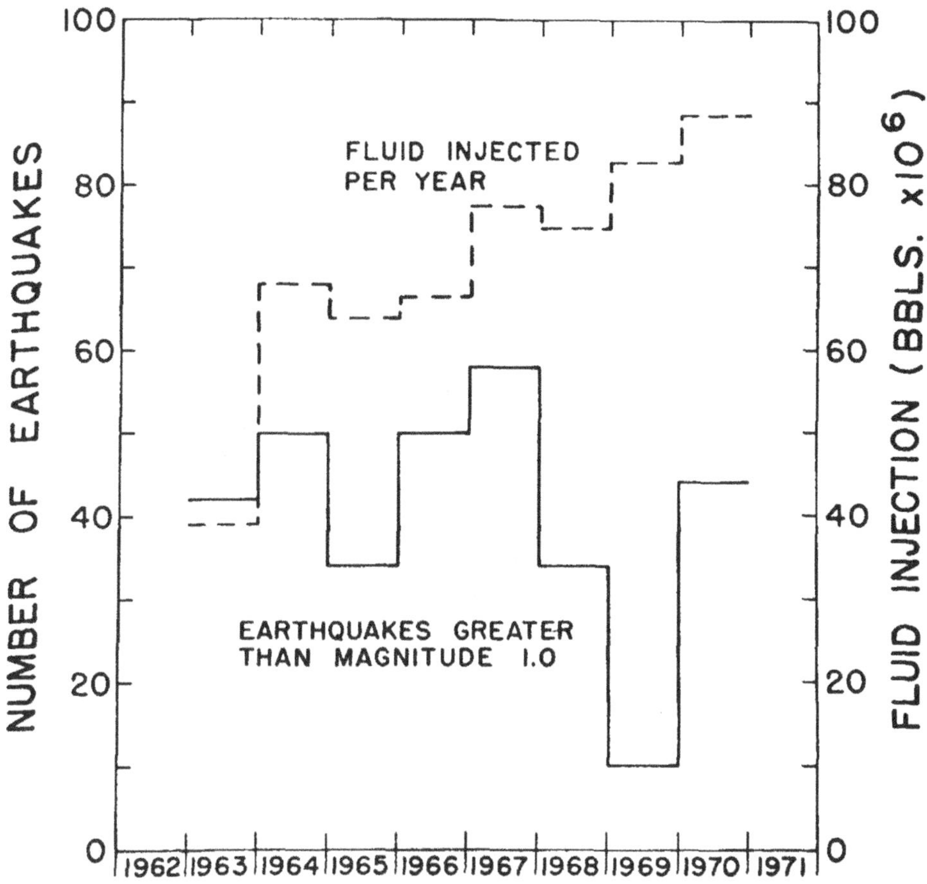

FIGURE 4. Total yearly fluid injection and number of earthquakes per year occurring in the Rangely oil field, Colorado and vicinity. (From Gibbs, J. F., Healy, J. H., Raleigh, C. B., and Coakely, J., *Bull. Seismol. Soc. Am.*, 63, 1569, 1973. With permission.)

pressure exceeded hydrostatic pressure or the original fluid pressure in the field[913,915] (Figure 5).

Figure 6 graphically displays the relationship between monthly reservoir pressure and earthquake frequency. In the fall of 1970, water was withdrawn from the reservoir for 6 months, allowing the pressure to drop rapidly from a high of 275 bars to about 170 bars. Within 1 month after the pressure reduction program commenced, the number of earthquakes dropped from 30 to 40 events per month to no more than 2 events per month[913] (Figure 6).

In late September 1971, the pattern of waterflooding in the field was changed by Chevron to increase oil production. Consequently, fluid pressures in the test area declined gradually until August 1972[915] (Figure 6). It is noted in Figure 6 that very few earthquakes occurred within 1 km of the four experimental wells during this period of declining pressure.

In 1972, Raleigh et al.[915,916] developed a computer model that used *in situ* stress and tensile strength measurements (field and laboratory) of the reservoir rock to predict the fluid pressure needed to trigger the earthquakes. Raleigh[910] has described the components of this research:

In the field, the absolute stresses in the reservoir rock and the orientation of the slip direction were measured.

FIGURE 5. Earthquake epicenters at the Rangely oil field, Colorado between October 1969 and November 1970. The heavy dashed line is a subsurface strike-slip fault. Contours are bottom hole 3-day shut in pressures (per square inch) as of September 1969 and are typical of recent pressure distributions in the field.[915] Triangles represent seismic stations and the rectangle represents the well used to measure *in situ* stress. The four injection wells used in the earthquake control experiment are located between the field boundary and the mapped fault. (From U.S. Geological Survey, *Earthquake Inf. Bull.*, 5, 7, 1973.)

We did this by the hydraulic-fracturing technique — that is, fluid pressure in a borehole was increased until the tensile stresses in the wall of the hole exceeded the tensile strength of the rock. At this point a tensile fracture opened and the pressure dropped as fluid seeped into the propagating fracture.

In the laboratory we calculated the tensile strength of the reservoir sandstone by measuring the breakdown pressure under known stresses and different rates of pressurization. Combining this with information on the fault and slip direction determined from nearby earthquakes, we obtained a value of 257 bars . . . as the critical fluid pressure required in the reservoir rock before earthquakes would be triggered.

In August 1970, a large booster pump was installed to raise the fluid pressure in the test area. Between October 1972 and January 1973, the bottom-hole pressure exceeded the predicted critical value of 257 bars (Figure 6), and the monthly average of seismic events within 1 km of the four wells increased to 6 (Figure 6). However, for February,

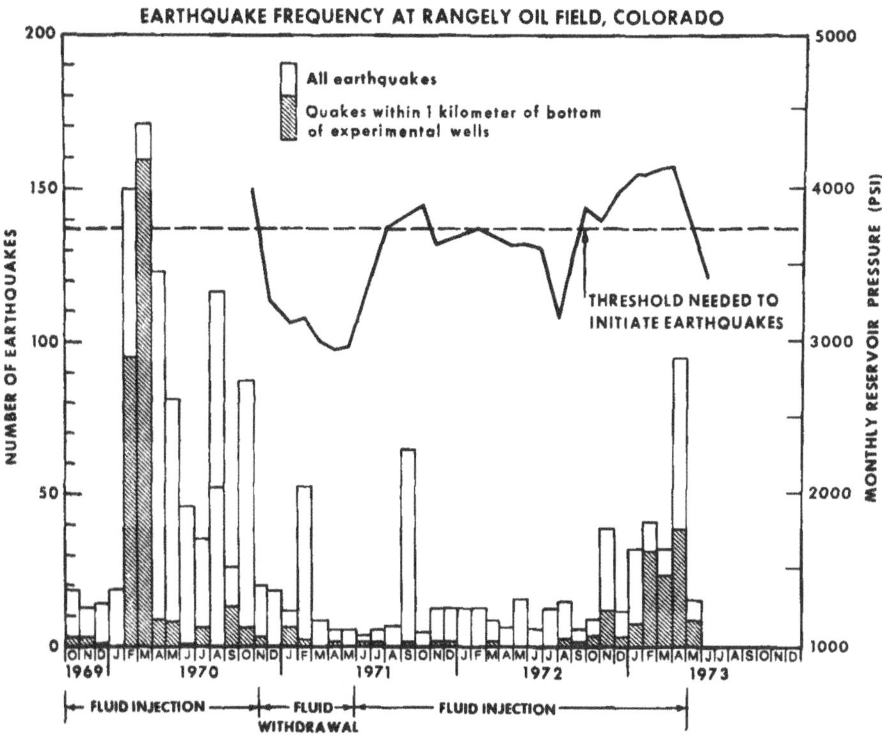

FIGURE 6. Earthquake frequency compared with monthly reservoir pressure at the Rangely oil field, Colorado. To convert reservoir pressure from pounds per square inch to bars divide by 14.7. (From U.S. Geological Survey, *Earthquake Inf. Bull.*, 5, 9, 1973.)

March, and April, earthquake frequency in the vicinity of the wells rose to about 30 events per month; during this 3-month period, the pressure stood at approximately 280 bars[915] (Figure 6). The prediction that a bottom-hole pressure of 257 bars was needed to initiate the seismic activity was borne out to within about 20 bars, or closer to 277 bars.[913] On May 6, 1973, the four wells were shut in and backflowing was begun. Since then, no events have occurred within 1 km of the wells, and only one event per month has been recorded along the fault zone to the southwest.[915]

The Rangely experiment clearly demonstrated for the first time that the occurrence of earthquakes could be controlled by manipulating the internal fluid pressure along an active fault. This discovery was extremely important because it means that a fault segment, at least in the shallow crust, can be weakened when the fluid pressure exceeds the predicted value or strengthened by reducing the internal fluid pressure. The former condition is accomplished by *fluid injection* and the latter by *fluid withdrawal.*

Reducing the internal fluid pressure creates a barrier to the propagation of a rupture, thereby making it possible to control the magnitude of earthquakes; the smaller the rupture length, the smaller the earthquake. At Rangely, the extraction of oil and water just to the northeast of the four experimental wells served to maintain the fluid pressure over most of the fault zone at values below the critical level needed for triggering earthquakes. Therefore, only a small segment of the fault was liable to shear failure.[916] Consequently, only small shocks were experienced along the fault zone near the four wells, and no events occurred along the fault northeast of the wells.

J. H. Dieterich and C. B. Raleigh[913,917] of the USGS have scaled down the Rangely oil field experiment to laboratory experiments in earthquake control. By subjecting a block of Westerly granite with a 45° saw cut (oriented to loading directions) to biaxial loading and varying fluid pressures along the slip surface with a system of injection and drain holes, they have demonstrated experimentally that it is possible to control the amount of slip occurring along the fracture.

C. Matsushiro, Japan Earthquakes

To determine if earthquakes could be induced by increased fluid pressure, the Japanese National Research Center for Disaster Prevention drilled an 1800-m deep well on the northeastern side of the Matsushiro strike-slip fault that was formed during the famous 1965 to 1967 swarm sequence (described in Volume I, Chapter 2) for a planned injection experiment.[918] It is likely that the bore encountered the fault at a depth of 600 m and was then guided by the fault plane.

During two injection periods in January to February 1970, water was pumped into the well at 50- and 14-bar pressures. Sudden increases in the daily number of small events followed increases in the well's water pressure with time lags of 9.3 and 4.8 days. Hypocenters (1) were distributed along a dipping plane that coincided with the fault and (2) migrated to deeper depths with time. According to Ohtake,[918] these patterns can be explained by the concept that a portion of the injected water permeated the Matsushiro fault, triggering the earthquakes.

D. Dale, New York and Los Angeles Earthquakes

Since August 1970, scientists from the Lamont-Doherty Geological Observatory of Columbia University have monitored the seismic activity in western New York.[919,920] During the first year of monitoring, the level of activity was less than one event per month near the community of Dale. However, between August and November 1971 when high pressures were attained in a hydraulic mining operation at Dale for the recovery of salt, the seismic frequency increased to as many as 80 events per day. The top-hole pressure of the water injection well was 120 bars. Epicenters were near the well and on or close to the Clarendon-Linden fault system. The seismic activity dropped to the preinjection frequency within 48 hr after the well was permanently closed in November 1971.

In August 1972, a second injection well located 0.5 km from the first well was put into operation; however, only a few earthquakes occurred near this well. Two similarities were noted for both wells: their hydrofracture and pressure histories were very similar, and they bottomed near the Clarendon-Linden fault system. However, the 1971 well was hydrofractured near the base of the salt layer, while the 1972 well was hydrofractured near the middle of the salt layer. The water loss was appreciable in 1971, but negligible in 1972. According to Fletcher et al.,[919] the differences in earthquake activity and water loss "are consistent with the hypothesis that fluids and thus pore pressures were confined to the salt layer in 1972, but those in 1971 penetrated into the fault zone in the rock unit below the salt layer."

Teng et al.[154] of the University of Southern California have conducted a monitoring program along a segment of the Newport-Inglewood fault system in Los Angeles to determine the relationship between waterflooding in the Inglewood oil field and microearthquake occurrence. For 1971, a correlation was suggestive (Figure 7), but as pointed out by Teng et al., "the present data do not yet permit a definite statement regarding the casuality of earthquake occurrence and oil field activities."

IV. FUTURE PROSPECTS

The success of the Rangely oil field experiment has suggested the notion that it might

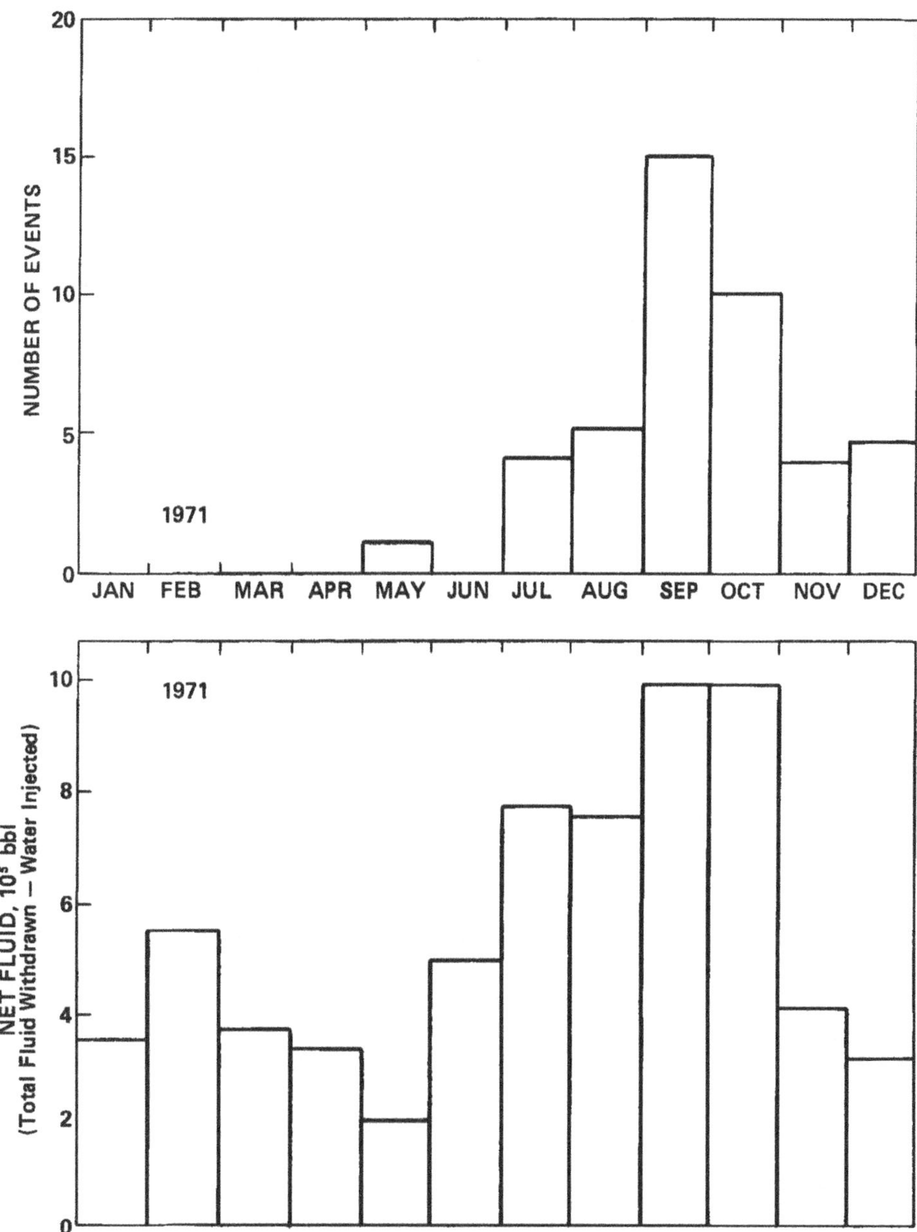

FIGURE 7. Correlation of earthquake occurrence and the net amount of fluid replacement, Inglewood, California oil field. (From Teng, T. L., Real, C. R., and Henyey, T. L., *Bull. Seismol. Soc. Am.*, 63, 874, 1973. With permission.)

be feasible one day to control earthquakes along fault zones that pose a hazard to urban areas. According to Raleigh et al.,[915] various schemes can be conceived that would lead to the size reduction of seismic events. For example, to accommodate slip along the San Andreas fault (which has a yearly slip rate of 2 to 3 cm)[921] without permitting sufficient tectonic strain to accumulate for yielding great earthquakes every 100 to 200 years, a M_L = 4.5 event would require a fault length of about 5 km and a slip of about 2 cm. To achieve these parameters, the following hypothetical scheme for the San Andreas fault has been developed by Raleigh et al. (Figure 8):

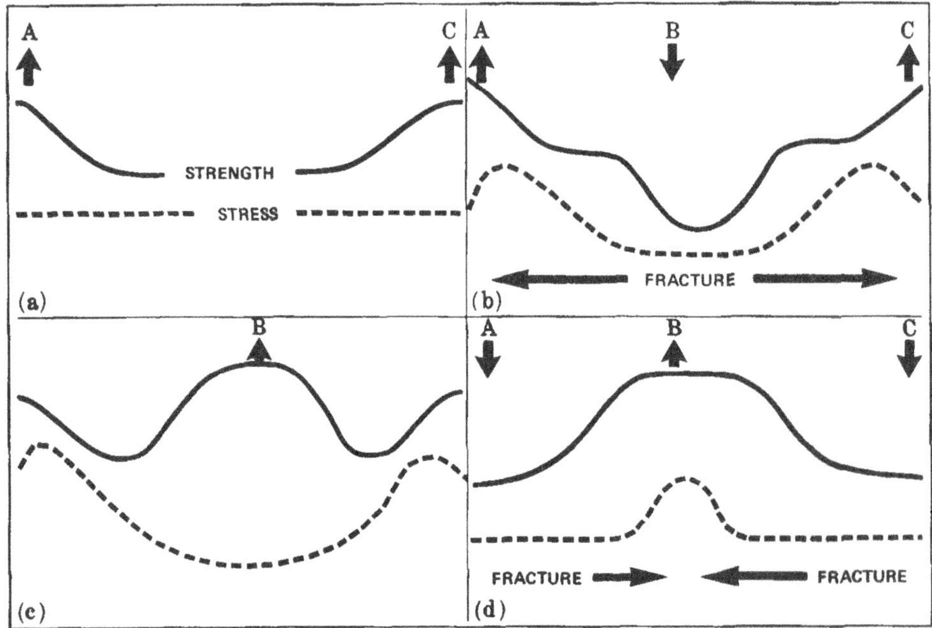

FIGURE 8. Hypothetical scheme for controlling earthquakes. (a) Fluid is removed from wells A and C, with an increase in frictional strength along the fault. (b) Fluid is injected into well B, triggering an earthquake; the stress drops at B and increases at A and C, where the fracture is arrested. (c) Fluid is removed from well B, resulting in an increase in strength at B. (d) Fluid is injected at A and C, producing earthquakes and the fracture is now arrested at B. (From Raleigh, C. B., Healy, J. H., and Bredehoeft, J. D., *Science*, 191, 1236, 1976. Copyright © 1976 by the American Association for the Advancement of Science. With permission.)

1. Wells about 5 km apart and 5 km deep are drilled along the fault zone, and the fluid pressure is reduced by the required amount and over a sufficient area of the fault to arrest a fracture of this size (Figure 8a).
2. Another well is drilled in the center of each 5-km segment, and fluid is then injected to trigger the earthquake (Figure 8b).
3. Within the faulted area, the stress is relieved, but stress becomes concentrated at the ends of the fracture in the strengthened zones. Fluid is then withdrawn from the wells formerly used for injection; this zone now becomes strengthened (Figure 8c).
4. Fluid is then injected into the intervening wells, and new seismic events are triggered at the former barriers (Figure 8d).

The above procedure would be alternated at 6-month intervals to accommodate the slip rate.

According to Raleigh,[910] the feasibility of this scheme "depends on factors about which we know very little." One of these is the permeability of the fault zone. If it turned out that the rocks in the fault zone were characterized by a very low permeability, it would take a large number of wells to render the zone effective as a barrier to rupture propagation. Presently, there are no data on permeability, temperature, level of stress, pore pressure, or the material properties of the fault zone at depths where earthquakes occur.[910,915] Raleigh[910] notes that these data could be provided if a number of holes approximately 7 km deep were drilled along sections of the San Andreas fault. Such a scheme has been proposed as a part of a deep drilling project of the continental crust by the U.S. Geodynamics Committee of the National Academy of Sciences.[922]

The future role of fluid injection as an agent to control earthquakes was summarized by Raleigh[923] in a presentation made to the Subcommittee on Oceans and Atmosphere, 92nd Congress:

> We believe that the results of the experiment at Rangely, Colorado, demonstrate the feasibility of earthquake control under certain circumstances. The question that we now face is how to apply this knowledge to active faults in the western United States.
>
> Seismic activity along these faults shows that most of the earthquake activity occurs at depths shallower than 10 km and practically no earthquakes take place at depths greater than 15 km. This shallow depth of seismic activity offers the possibility of drilling into active seismic zones and attempting to control the activity by controlling the fluid pressure. There are many strategies for modifying seismic behavior along a fault without creating a large earthquake. All of these involve the concept of first locking portions of the fault so that there is no possibility of a great earthquake resulting from the control efforts, and then allowing controlled slip to take place between the locked points by increasing the fluid pressure.
>
> No one at this time can predict whether practical earthquake control will be feasible. Before any reasoned judgment on this matter can be reached, it will be necessary to have detailed information about the physical properties within the fault zone. Such information can only be obtained by drilling a number of holes along the fault to be controlled.

Indeed, much work still remains, and according to Healy et al.,[924] the USGS is searching for an unpopulated site along an active fault for conducting additional control experiments.

Chapter 3

EARTHQUAKE-RESISTANT PROVISIONS FOR STRUCTURES

I. INTRODUCTION

Because most earthquake-related deaths and injuries are caused by the partial or total collapse of structures (e.g., buildings, towers, chimneys), society has determined that the primary mitigation measure to guard against the effects of an earthquake is to control the standards of construction in areas having a seismic risk. The standards of construction are regulated by the earthquake-resistant regulations of *building codes*.[925]

In the latest edition of *Earthquake Resistant Regulations: A World List 1973*,[926] 27 countries were reported to have some type of published seismic regulations.* The majority of these building code regulations are outmoded and in need of major improvements. An official design code for earthquake-resistant industrial and public structures in the People's Republic of China was completed by the Ministry of Construction in 1974.[873] Unfortunately, many countries with a long history of devasting earthquakes have no seismic building codes or ones that have never been implemented. According to Dowrick,[927] there are more than 60 countries where earthquake forces should be considered in the design of structures. It should be noted that, although a country does not have an official earthquake-resistant code, many modern buildings may have been designed and constructed in accordance with a code used in another country. This is determined by the owner of a specific structure or perhaps by a lending institution.

The Committee on Seismology, National Academy of Sciences,[928] has summarized the general principles that govern the preparation of an earthquake-resistant building code.

> The code is not a technical treatise on earthquake-resistant design, but is a concise statement of currently accepted professional practice. It does not attempt to relieve the individual structural engineer of the need to exercise a high degree of judgment in design details. In general, the code is more an expression of desired results that a set of instructions as to how to attain them.
>
> Earthquake-resistant-design codes are not intended to insure against damage to structures. It is assumed that large earthquakes will cause heavy damage..., but it is intended that they will not cause building collapse with consequent loss of life and injury. The code thus contains an implicit economic judgment as to a reasonable balance between repair costs and initial costs. Since such judgments will depend very much on local conditions, it would be expected that different countries and different areas in the same country might well have very different codes.
>
> All codes are in a constant state of development and improvement. As new research knowledge becomes available, as experience from destructive earthquakes accumulates, and as various social and economic changes appear, it becomes necessary to modify the code. A reasonable degree of flexibility in formulation, interpretation, implementation, and revision thus becomes important.
>
> All experience has convincingly demonstrated that codes themselves are of little use unless they are backed by a powerful enforcement agency and a comprehensive inspection service.

II. LATERAL EARTHQUAKE FORCES

Ground motion is the principal force mechanism addressed by the seismic provisions of a building code. This is the earthquake hazard usually responsible for the greatest

* These were Algeria, Argentina, Austria, Bulgaria, Canada, Chile, Cuba, El Salvador, Federal Republic of Germany, France, Greece, India, Iran (not official), Italy, Japan, Mexico, New Zealand, Peru, Philippines, Portugal, Rumania, Soviet Union, Spain, Turkey, U.S., Venezuela, and Yugoslavia.

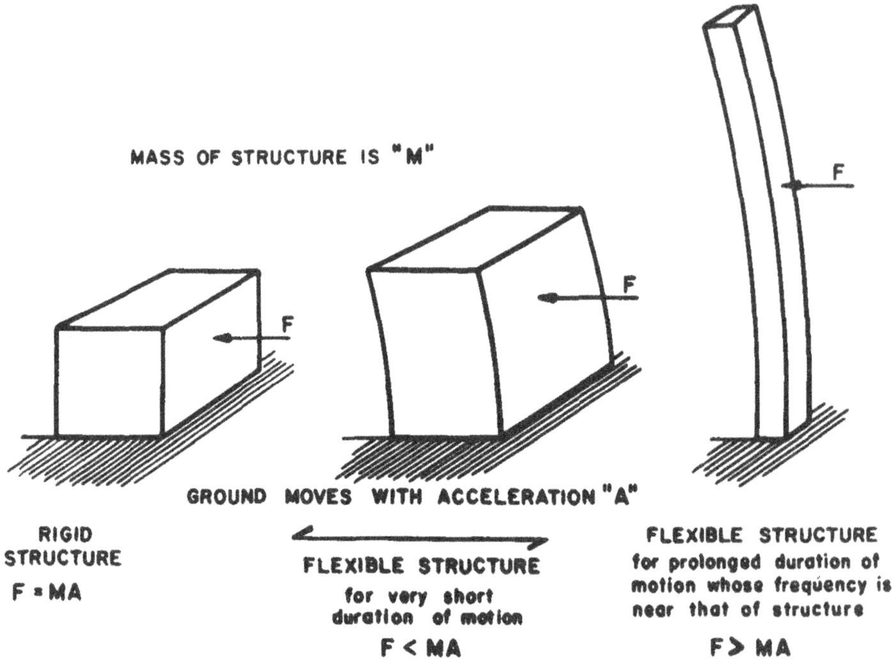

FIGURE 1. Earthquake-induced ground motion setting up *inertia forces* in three types of buildings. F = inertia forces, M = mass of the structure, and A = acceleration of the ground motion at any instant. (From Degenkolb, H. J., Earthquake Forces on Tall Structures, Booklet 2717A, Bethlehem Steel Corporation, Bethlehem, Pa., 1977, 3. With permission.)

losses to the works of construction. Violent and erratic ground motions, both with horizontal and vertical components, set-up complex inertia forces (i.e., the *earthquake force*) in overlying structures. The earthquake forces are *dynamic*, not *static*, because they are time variant; that is, the "loading of all aspects of the structure response (deflections, internal force, stress, etc.) all vary with time."[929]

The primary purpose of a seismic building code is to provide design and construction standards that will enable a structure to *resist* the horizontal or lateral forces of an earthquake without major failures. Because a structure must be able to withstand its own weight (i.e., *dead load*) plus the weight of its contents (i.e., *live load*) there is a built-in resistance to the vertical component of the ground motion.[930] Consequently, compensating measures for this type of dynamic load usually are not considered in earthquake-resistant provisions.

A. Types of Structures and Structural Materials

In order for a structure to resist horizontal seismic loads, it must incorporate the appropriate flexibility and energy-absorption capacity to insure that displacements will take place, but without large inertia forces being produced within the structure.[931] Degenkolb[932] has shown how inertia forces can differ for three different types of buildings (Figure 1). For a building that is rigid and firmly coupled to its foundation, the inertia forces (F) are equal to the mass of the structure (M) times the acceleration of the ground motion (A). For a building that yields slightly and is exposed to a short duration of ground motion, the inertia forces are somewhat smaller than the product MA because the flexing of the structure absorbs part of the energy. The inertia forces may be considerably larger than MA if a building is very flexible and is subjected to several cycles of ground motion with a period approximating the natural period of the building.

Buildings designed specifically to resist lateral earthquake forces usually have a *fully flexible frame*, a *modified flexible frame*, or a *composite frame*. A fully flexible system will bend or flex when earthquake forces are applied. The bending takes force to accomplish; hence, the frame absorbs a portion of the energy imparted to the structure.[933] The frame consists of vertical *columns* and horizontal *beams* or *girders*. *Diagonal structural members*, such as X-braces, may or may not be used;[934] diagonal components are used to strengthen the columns. Nonstructural components of the building are fully separated from the frame allowing it to function as a complete, self-contained unit. Fully flexible structures with long natural periods are well suited for short-period foundation sites[927] (e.g., bedrock, firm soil).

A modified flexible system consists of a *stiff* or *"rigid" frame* with additional resistance provided by *shear walls*. Such walls are solid vertical partitions, usually reinforced concrete, that extend between the beams and columns of the frame; shear walls resist lateral forces parallel to the walls. The stiff frame/shear wall arrangement provides dual paths for the transfer of inertia forces to the ground, with the more flexible frame providing a second line of defense should the stiffer shear walls fail.[935] Modified flexible structures are especially suited for long-period foundation sites[927] (e.g., thick alluvium).

Some buildings are designed with a mixture of flexible and rigid frame components. One common type of composite frame is the *"open first floor" system* wherein a rigid upper structure is placed on a flexible column frame. The columns are expected to resist exaggerated and concentrated lateral forces.[933] This type of system has performed poorly in several recent earthquakes.

The capacity of a structure to absorb energy within certain limits of deformation and without failure is one of the most desirable characteristics of earthquake-resistant design. Because most building materials have only a limited energy-absorbing capacity in the elastic range of deformation, there is a need for a material to have a high energy absorption capacity in its inelastic range of deformation.[934] *Ductility* refers "to the ability of a material to absorb energy while undergoing deformation without failure, particularly when the direction of the forces involved changes several times."[933]

Structural steel and ductile reinforced concrete exhibit a high level of ductility and are the only materials that can be used for fully flexible high-rise structures. Wood also exhibits high ductility when used in low-rise buildings. As reported by Botsai et al.,[933] a ductile structural system "can be thought of as providing a quality of toughness which, to a large extent, determines a building's survival under seismic conditions." Many commonly used building materials have a low level of ductility because they exhibit very little inelastic behavior under loading and fail at or near the inelastic limit. Materials in this category include, typical reinforced concrete, precast concrete, brick, concrete block, and adobe — the so called *brittle materials*.[933,934]

The favorable ductile characteristics of a steel frame compared to a concrete frame have been described by Degenkolb[932] for a modified flexible building (Figure 2). Because the frame must act as a second line of defense, it must remain intact should the concrete shear walls fail. With a steel frame, the strength of the members is not affected by cracking in a shear wall; therefore, it can carry the required load after a shear wall has failed. By contrast, a crack passing through a shear wall can penetrate and severely weaken the concrete frame (Figure 2).

Wood-frame buildings normally have high ductility because of their small height and weight. However, proper bracing and connections to the foundation and proper workmanship are essential elements to insure structural integrity during seismic loading. For example, the frame can slide across or even off the foundation during moderate shaking if it is not securely anchored to the foundation.[27] The structural sound-

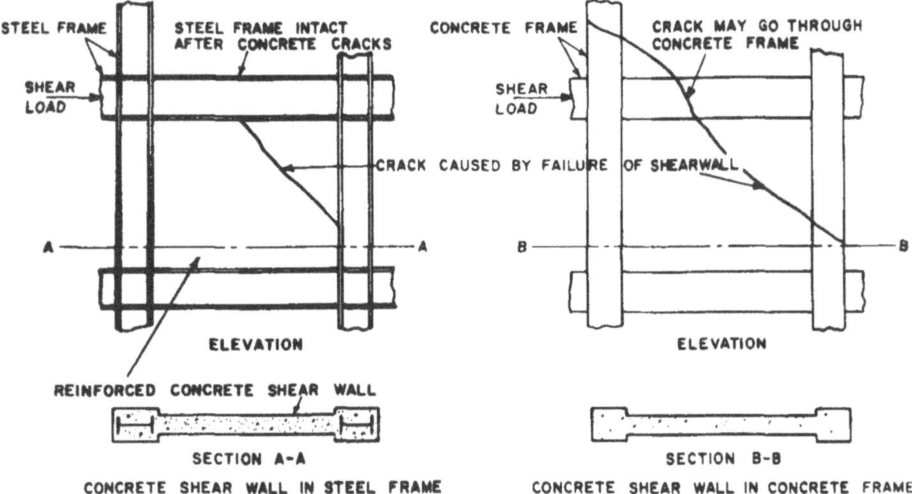

FIGURE 2. Advantage of using a *ductile material*, such as steel rather than concrete, for the basic frame of tall structures. Note that failure in a concrete *shear wall* can penetrate and damage the concrete frame. (From Degenkolb, H. J., Earthquake Forces on Tall Structures, Booklet 2717A, Bethlehem Steel Corporation, Bethlehem, Pa., 1977, 23. With permission.)

ness of a wood-frame building usually deteriorates with age. Examples of structural deterioration are rotting due to damp ground and termite damage.

B. Important Principles for Safe Building Design

Adhering to several design principles can contribute to the favorable response of a building subjected to earthquake forces. In general, a building should have (1) *a simple plan and elevation configuration,* (2) *equal stiffness throughout,* and (3) *sound structural connections.*

One of the most important elements of an earthquake-resistant structure is its plan and elevation configuration. The optimum choice is a symmetrical configuration, such as a square cube, because the induced vibrations and displacements will be uniform or nearly so during an earthquake. However, it is usually not possible to satisfy the demand for a symmetrical building because of lot size and shape restrictions and functional requirements of the proposed structure. Consequently, it is important to know how variations in symmetry can affect the performance of a building subjected to seismic loading.[927]

L, T, or U shapes are often adopted because of the above constraints. For such irregular shapes, it must be realized by the structural engineer that the wings can experience different movements dependent upon their orientation to the direction of the earthquake force.[927] Consequently, the *reentrant corners* (i.e., corners where the wings intersect) are areas of high stress concentration, and they must be reinforced accordingly to resist potential overstresses[932] (Figure 3).

An extremely long building in plan should be avoided because different earthquake motions may be applied simultaneously to both ends of the structure resulting in unequal vibrations and deflections. This problem is aggravated if the building transverses ground with different vibration characteristics. The best solution for an elongated building is to design two totally separate buildings.[927]

Two important principles apply to elevation configurations. First, if a building is too slender it can experience excessive horizontal deflections under seismic loading, creating potential overstresses in the outer columns.[927] Dowrick[927] suggests a limited

FIGURE 3. High stress concentrations at the *reentrant corners* of three irregular shaped buildings in plan. (From Degenkolb, H. J., Earthquake Forces on Tall Structures, Booklet 2717A, Bethlehem Steel Corporation, Bethlehem, Pa., 1977, 19. With permission.)

slenderness of height/width $\leqslant$ 3 or 4 for most buildings. Second, the floor area of upper stories should not be larger than those below. This produces a top-heavy building that can lead to exaggerated deflections because the structure is analogous to an inverted pendulum.

Buildings with irregular shapes are especially vulnerable to *torsion* (i.e., twisting or rotation about an axis) due to an eccentricity between the center of mass and the center of rigidity. For a cross-shaped building like the one depicted in Figure 4, the A-bents, by reason of their lengths, will take more lateral load than the shorter B-bents. However, if the building twists because of earthquake torsion, the A-bents are inefficient due to their short movement arm, and much of the torsional load must be carried by the B-bents. Therefore, the B-bents should be designed for more torsional loading than otherwise would be expected from bent rigidities if torsion is ignored.[932]

Torsion also can be a problem in regular-shaped buildings if the relative stiffness is unevenly distributed. For example, in a rectangular building with a very rigid, off-center core and the remainder of the structure flexible, torsion can develop in the flexible portion around the stiffer core[933] (Figure 5). Torsion can be a problem in commercial buildings that have an open front in the first floor for displaying merchandise. It is impossible to design this wall to equal the strength and stiffness of the other walls. Consequently, the building is apt to twist during an earthquake, amplifying the deflections at the front wall.[936]

Regardless of the structural system used to resist seismic loading, it must respond as a coherent unit. The basic essential is *unit strength* in which all structural assemblages are securely tied together. If this is not done, separate structural elements will respond individually to the earthquake force, and failure will commence at the weakest component. With failure, there is a shift in the load carrying or resisting ability to the other elements, which in turn can fail due to overloading.[933] Connection schemes need special attention in masonry construction. As described by Botsai et al.,[933] when the floors are not properly attached to the walls, the units move independently during seismic loading. This can cause the walls to fail or the floors to drop (Figure 6).

III. U.S. SEISMIC CODE PROVISIONS

Because of its wide usage, the "recognized" building code in the U.S. is the International Conference of Building Official's (ICBO) *Uniform Building Code* (*UBC*), which "covers the fire, life and structural aspects of all buildings and other related structures."[937] Local jurisdictions can enact the UBC as written, or any number of the provisions can be amended, rewritten, or deleted to better reflect special problems or needs of a particular community. In addition, a local civic authority may augment the UBC with supplemental building ordinances. A jurisdiction need not adopt the most

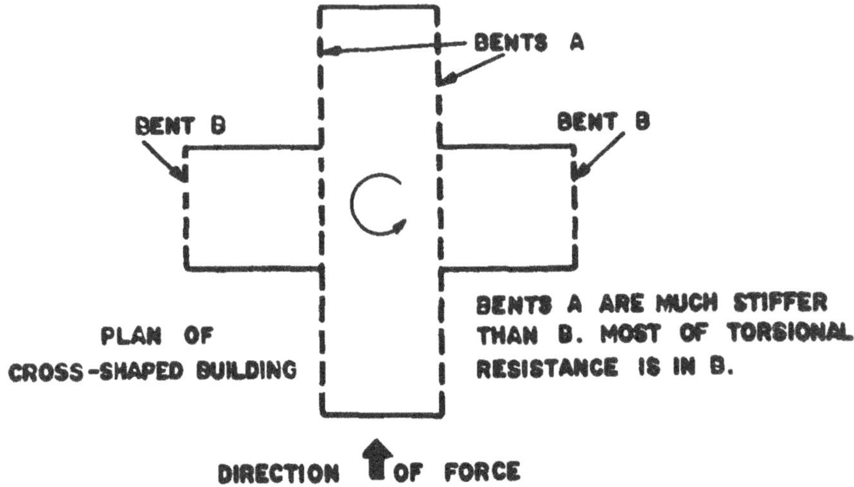

FIGURE 4. *Torsional resistance* for a *cross-shaped building* in plan. (From Degenkolb, H. J., Earthquake Forces on Tall Structures, Booklet 2717A, Bethlehem Steel Corporation, Bethlehem, Pa., 1977, 19. With permission.)

recent edition of the UBC, although it is usually thoroughly revised at 3-year intervals. For example, the City of Long Beach, California uses the 1970 and not the 1976 (latest) edition of the UBC. Large civic enities may write their own building code, as in the case of Los Angeles[938] and San Francisco,[939] with the UBC serving as a model.

Most of the earthquake-resistant provisions are contained in Chapter 23, Section 2312 of the UBC. This entire section is presented in Appendix D. The earthquake provisions of the UBC and the building codes of Los Angeles and San Francisco are based upon and closely follow the Structural Engineers Association of California's (SEAOC) *Recommended Lateral Force Requirements,* which provide minimum design criteria in broad general terms for the construction of earthquake-resistant structures. The SEAOC recommendations are supplemented by the SEAOC *Commentary* which elaborates on the recommended requirements, explains their intent, and acts as a guide for the application of the recommendations.[934]

Because the primary function of the earthquake provisions of the UBC is to guard against major failures, more specialized regulations to resist earthquake forces have been implemented by various jurisdictions to cover certain aspects of construction and unique types of structures. In addition, special ordinances have been implemented to restrict the building of certain types of structures on or near active fault zones. For example, the State of California requires special seismic standards for public school and hospital buildings, and the U.S. Nuclear Regulatory Commission enforces provisions that apply to the site selection and construction of nuclear power plants. Several of these special provisions are described later in this chapter.

A. Historical Development of Seismic Regulations

Seismic regulations for buildings in the U.S. were founded in California. The impetus was provided, in large measure, by the widespread destruction to buildings caused by the Santa Barbara earthquake of June 29, 1925 (M_L = 6.3) and the Long Beach earthquake of March 10, 1933 (M_L = 6.3).

During the rebuilding efforts that followed the April 18, 1906 San Francisco earthquake (M_S = 8.3), building codes called for a lateral force resistance of 30 lb/ft^2 (psf) for *wind loads.* The term "earthquake" was apparently never mentioned in the codes.

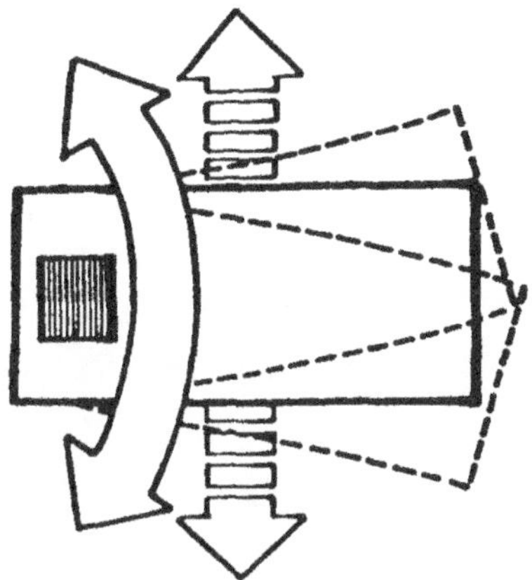

FIGURE 5. *Torsion effect* in a rectangular building with a very rigid off-center core and a flexible outer (surrounding) structure. (From Botsai, E. E., Goldberg, A., Fisher, J. L., Lagorio, H. J., and Wosser, T. D., *Architects and Earthquakes*, AIA Research Corporation, Washington, D.C., 1977, 42. With permission.)

The wind load was later reduced to 20 psf in 1910 and 15 psf in 1920. However, during the ensuing years, leading structural engineers in California introduced the Newtonian concept of *lateral earthquake forces proportional to the weight of the structure* (i.e., the lateral forces will act on a structure in proportion to its weight). This concept first found its way into building codes in 1927 when it became a suggested provision in the first edition of the Uniform Building Code. Basically, the code called for a lateral earthquake force of 10% of the total dead and live loads of the proposed structure when the soil pressure was equal to or greater than 2 tons/ft^2. The lateral force was 7.5% of the total dead and live loads when the soil pressure was less.[940] This would mean that structures covered by the code had to be able to resist a constant lateral force equal to either 10% or 7.5% of their own weight, depending upon the character of the foundation material. This relation can be expressed as:

$$V = CW \tag{1}$$

where V = total lateral earthquake force, C = numerical coefficient for the horizontal force factor, and W = weight (dead + live loads) of the proposed structure.

Following the 1925 Santa Barbara earthquake, the City of Santa Barbara adopted earthquake provisions in its municipal code, and this represented the first code in California to incorporate seismic provisions. Palo Alto adopted similar provisions a short time later.[934,940]

In 1928, the California State Chamber of Commerce called for the establishment of a building code that would be "dedicated to the safeguarding of buildings against earthquake disasters." The Joint Committee on Seismic Safety[940] has summarized the scope of the chamber's intent and influence on effecting earthquake design in California.

It is interesting to note that the chamber launched this project at the urgent request of business interests who were concerned about the sharp recession in building, the increased costs of earthquake insurance, and the amount of such insurance required by the State Corporation Commissioner before he would approve

FIGURE 6. Reinforced concrete Mansion Charaima Apartment building in Caracas, Venezuela following the July 29, 1967 earthquake (M_L = 6.5). The top four floors "*pancaked*" causing 42 deaths.[933] (Courtesy of the National Oceanic and Atmospheric Administration, Environmental Data Service.)

bond issues on certain types of structures following the 1925 Santa Barbara earthquake. Studies made under the chamber's sponsorship included work by many of the State's leading structural engineers, architects, and building contractors, and resulted in a document entitled *Building Code for California,* which formed the foundation for the codes that followed. This comprehensive code document, covering structural design as well as fire and panic considerations was published in 1939.

In the same year as the destructive 1933 Long Beach earthquake, the California Legislature adopted the *Riley Act* and the *Field Act.* The Riley Act specified that all buildings, except certain dwellings and farm buildings, had to be designed to resist earthquake forces proportional to their masses. The initial regulations called for a building to be designed to resist a lateral force of 2% of its total vertical design load. This requirement was revised in 1953, requiring a lateral resistance of 3% for buildings under 12.2 m in height and 2% for those over 12.2 m in height. In 1965, the requirements were again revised by the legislature to conform with the seismic regulations of the Uniform Building Code.[940] The Field Act pertains to the seismic provisions for public school buildings and is discussed in a later section of this chapter.

In 1957, the Structural Engineers Association of California formed the statewide *Seismology Committee* to resolve differences in existing building codes and prepare a single set of lateral force requirements that would be acceptable to the state's structural engineers and that would be used for the design and construction of structures in the seismic areas of the U.S., but especially in California. The committee's "Recommended Lateral Force Requirements" were adopted by the SEAOC in 1959 and published with a Commentary in 1960.[934] Since 1959, the requirements and commentary have been refined and expanded several times, the latest being in 1974/1975. The SEAOC recommended requirements have been adopted in whole or in part by many code writing authorities throughout the world.

The seismic design intent or philosophy of the SEAOC Recommended Lateral Force Requirements is illustrated by the following quotation from the 1975 SEAOC Commentary.[934]

> The SEAOC Recommendations are intended to provide criteria to fulfill life safety concepts. It is emphasized that the recommended design levels are not directly comparable to recorded or estimated peak ground accelerations from earthquakes. They are, however, related to the effective peak accelerations to be expected in seismic events. More specifically with regard to earthquakes, structures designed in conformance with the provisions and principles set forth therein should, in general, be able to:
>
> 1. Resist minor earthquakes without damage;
> 2. Resist moderate earthquakes without structural damage, but with some nonstructural damage;
> 3. Resist major earthquakes, of the intensity of severity of the strongest experienced in California, without collapse, but with some structural as well as nonstructural damage.
>
> Conformance to the Recommendations does not constitute any kind of guarantee that significant structural damage would not occur in the event of a maximum intensity earthquake. While damage in the basic materials now qualified may be negligible or significant, repairable or virtually irrepairable, it is reasonable to expect that a well-planned structure will not collapse in a major earthquake. The protection of life is reasonably provided, but not with complete assurance.
>
> It is to be understood that damage due to earth slides such as those that occurred in Anchorage, Alaska, or due to earth consolidation such as occurred in Niigata, Japan, would not be prevented by conformance with these Recommendations. The SEAOC Recommendations have been prepared to provide minimum required resistance to typical earthquake ground shaking, without settlement, slides, subsidence, or faulting in the immediate vicinity of the structure.

The added expense to design and construct a building that provides minimum standards to resist ground motions (above three resistance levels) is 1 to 2% of the total project costs in most cases, and 2 to 10% in a minority of cases.[941]

IV. UNIFORM BUILDING CODE — LATERAL DESIGN PROVISIONS

Beginning with the first edition of the Uniform Building Code in 1927 and continuing until 1961, the earthquake provisions were placed in the appendix as suggested ordinances, enabling a jurisdiction to exclude their adoption. However, in 1961 the lateral force provisions were moved to the main body of the UBC, thereby becoming mandatory unless a jurisdiction specifically excluded their adoption and were revised to be in general agreement with those recommended by the SEAOC.[925,940] The following discussion is concerned with several of the lateral design regulations of the 1976 edition of the Uniform Building Code.[937] The entire "Earthquake Regulations" section of the UBC is presented in Appendix D of Volume III.

A. Minimum Earthquake Forces for Structures — Equivalent Static Analysis

The equivalent static analysis method, also known as the *seismic coefficient method,* is used for establishing minimum lateral earthquake forces. The analysis consists of reducing the dynamic seismic force into *static* or *at rest equivalents* and designing a structure to withstand these loads. This method is normally used for structures of low to medium height with symmetrical designs[930] (i.e., structures for which dynamic characteristics do not vary greatly).[927]

The *total lateral force or shear at the base of a structure* (V), acting nonconcurrently in the direction of each major axis, is determined by:

$$V = ZIKCSW \qquad (2)$$

where Z = *seismic risk zone factor,* I = *occupancy importance factor,* K = numerical

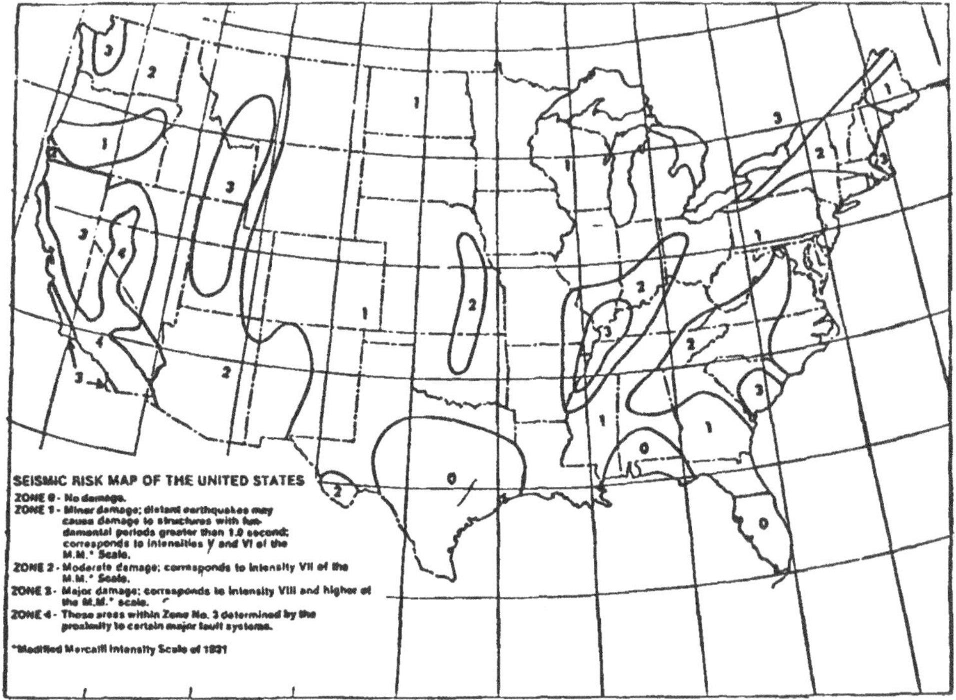

FIGURE 7. Seismic risk map of the conterminous U.S. used in the *Uniform Building Code — 1976 Edition.* (Reproduced from the 1976 edition of the Uniform Building Code, copyright © 1976. With permission of the International Conference of Building Officials, Whittier, Calif.)

coefficient for the horizontal force factor for various types of framing arrangements (i.e., *framing factor*), C = numerical coefficient for base shear (i.e., *flexibility factor*), S = numerical coefficient for site-structure resonance (i.e., *site factor*), and W = total dead load of a proposed structure and portions of other loads when applicable (i.e., *weight factor*).

According to the SEAOC,[934] the minimum design forces derived from Equation 2 are not to be considered as being the actual forces to be expected during an earthquake. Actual earthquake motions may be greater than the motions used to produce the prescribed minimum design forces. The justification for permitting lower force values for design include "increased strength beyond working stress levels, damping contributed by all the building elements, an increase in ductility by the ability of members to yield beyond elastic limits, and other redundant contributions."[934]

1. Z or Seismic Risk Zone Factor

Because the UBC is used throughout the U.S., the country is divided into zones of varying seismic risk (Figures 7 and 8). The criteria used to define these zones include: (1) the distribution and frequencies of known damaging earthquakes and the intensities associated with these events, (2) the evidence of energy release, (3) the geologic structures and provinces that are believed to be associated to earthquake activity, and (4) the proximity to certain major fault systems. The degrees of seismic risk are

- Zone 0 = no damage.
- Zone 1 = minor damage, but distant earthquakes may cause damage to structures with fundamental periods greater than 1.0 sec, Intensities V and VI on the

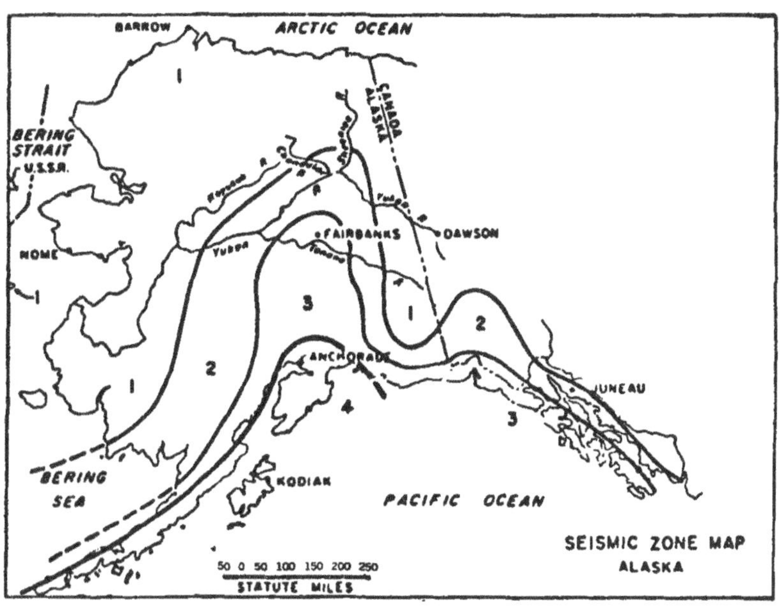

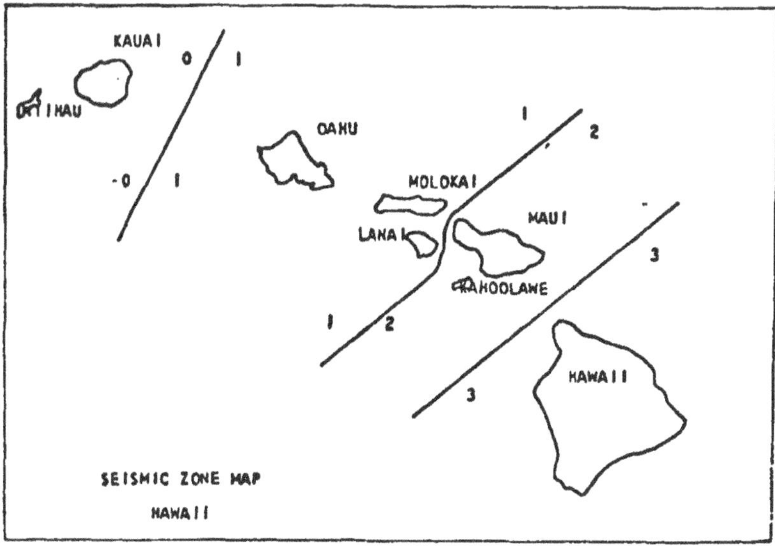

FIGURE 8. Seismic risk maps of Alaska and Hawaii used in the *Uniform Building Code — 1976 Edition.* (Reproduced from the 1976 edition of the Uniform Building Code, copyright © 1976. With permission of the International Conference of Building Officials, Whittier, Calif.)

Modified Mercalli Intensity Scale of 1931 (see Table 8 in Volume I, Chapter 2, for intensity level definitions).

- Zone 2 = moderate damage, Intensity VII.
- Zone 3 = major damage, Intensity VIII or higher.
- Zone 4 = areas within Zone 3 in close proximity to certain major fault systems.

Approximately one third of the population is located in Zones 2, 3, and 4.[925]

Zone 4 was used for the first time in the 1976 edition of the UBC and delineates the areas of highest seismicity in California, Nevada, and Alaska. The other zones do not account for variations in seismicity. The distances from the fault systems and the potential Richter magnitudes that set the boundaries of Zone 4 are[934]

Potential Richter magnitude	Distance from fault
7.0 or greater	40.3 km (25 mi)
6.0 to less than 7.0	24.2 km (15 mi)

The Z values for Equation 2 are 3/16, 3/8, 3/4, and 1.0 for Seismic Risk Zones 1, 2, 3, and 4, respectively. Therefore, as the zone factor increases, all else remaining equal, a structure must be designed and constructed to resist progressively larger lateral forces. No risk zone variations are used in the Los Angeles and San Francisco building codes; the risk factor is considered to be unity or 1.0.

2. I or Occupancy Importance Factor

The occupancy importance factor was used for the first time in the 1976 edition of the UBC and is an outgrowth of the February 9, 1971 San Fernando earthquake (M_s = 6.5). As a result of this seismic event, heavy damage was sustained "to facilities deemed essential to the public welfare during postearthquake operations."[934] The unsatisfactory performance of several hospital buildings in the San Fernando Valley is discussed later in this chapter.

Values for the I factor are 1.5 for essential facilities, 1.25 for any building where the primary occupancy is for the assembly of more than 300 persons, and 1.0 for all other buildings (Table 23-K, Appendix D). Essential facilities shall include hospitals, fire and police stations, and governmental disaster operation and communication centers. A jurisdiction can expand the essential facility list to include other buildings that it deems necessary for emergency uses following an earthquake.

3. K or Framing Factor

The purpose of the K factor is "to give all types of structures an equal probability of performance under a designated earthquake."[934] Structures that have high ductility or inherent resistance and that have performed well in past earthquakes have been assigned lower K values by the Seismology Committee of the SEAOC, while those that have performed poorly have been assigned higher K values.

For buildings, K values range from 0.67 to 1.33 (Table 23-I, Appendix D). K = 0.67 is reserved for a ductile, moment resisting frame. Such a frame has the capacity to resist the total required lateral force without reliance on shear walls or other forms of rigid bracing. It is composed of structural steel or ductile reinforced concrete. K = 1.33 is used for buildings with a box system in which the required lateral forces are resisted by shear walls or braced frames.

Structures other than buildings have higher K values, usually 2.0, for reasons as given by the Seismology Committee.[934]

> This is justified in relation to the K values stipulated for buildings (0.67 to 1.33) on the basis that most of these other structures do not have the multiplicity of structural and non-structural resisting elements characteristic of most buildings; do not have a significant natural damping; do not have elements which could be permitted to yield or even fail without jeopardizing the safety of the structure. While these structures do not generally constitute as much of a personal hazard as do buildings with their high human occupancy, they frequently do represent a significant property loss risk, offsetting their lack of personal hazard. Taken all together, it is felt that the higher factor for these structures is justified.

Elevated, cross-braced tanks carry a K value of 2.5, the highest (poorest) possible rating, because of their undesirable dynamic characteristics (acting similar to an inverted pendulum) during earthquake ground shaking (Figure 9) and the importance for maintaining their integrity after an earthquake for possible fire-fighting efforts.[934]

4. C or Flexibility Factor

The flexibility factor is dependent upon a proposed structure's period of vibration and is determined by:

$$C = \frac{1}{15\sqrt{T}} \tag{3}$$

where T = fundamental elastic period of vibration in seconds in the direction under consideration. Three formulas are available for determining a proposed structure's period of vibration (Formulas 12-3, 12-3A, 12-3B; Appendix D).

Equation 3 has undergone several revisions. In 1943, the City of Los Angeles recognized indirectly the influence of a structure's flexibility on earthquake design coefficients by revising the values to reflect the number of stories in a building.[940] The value of C was determined by:

$$C = \frac{60}{N + 4.5} \tag{4}$$

where N = total number of stories above the one under consideration. This formula was adopted when buildings in Los Angeles could not exceed 13 stories (48 m), and hence, it was never intended to be used for structures exceeding this height.

In 1959, the height limitation was removed in Los Angeles, and Equation 4 became:

$$C = \frac{4.6\,S}{N + 0.9(S - 8)} \tag{5}$$

where S = total number of stories in the structure for buildings exceeding 13 stories. S = 13 was adopted for all buildings less than 13 stories in height.

In order to update the C coefficient in the San Francisco Building Code, the *Joint Committee on Lateral Forces* was established in 1948 by the San Francisco section of the American Society of Civil Engineers and the Structural Engineers Association of Northern California.[942] After several years of study, the committee recommended C coefficients that were related to the calculated period of the structure in accordance with:

$$C = \frac{K}{T} \tag{6}$$

where K = 0.15 for buildings and 0.025 for other structures. For buildings, C_{max} = 0.06, C_{min} = 0.02; for other structures, C_{max} = 0.10, C_{min} = 0.03. The C coefficients were normally applied to the dead load plus 25% of the live load to arrive at the lateral earthquake force. In the case of warehouse and storage occupancies, 50% of the live load was added to the dead load and multiplied by C. The relation in Equation 6 was one of the first attempts to represent the dynamic behavior of structures in earthquakes.[943] San Francisco adopted Equation 6 in 1956, but with K = 0.02 for buildings and K = 0.035 for other structures.

In the 1973 edition of the UBC,

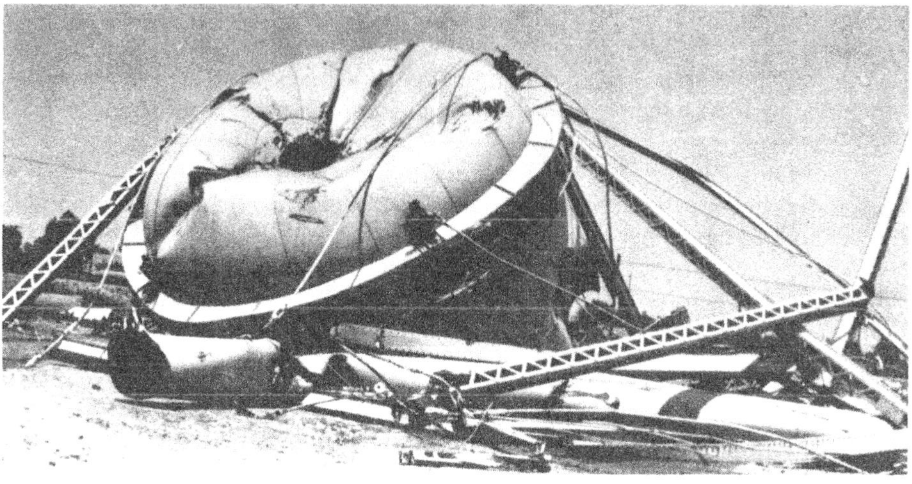

FIGURE 9. Crushed 383-m³ steel water tank and 30-m steel tower in Bakersfield, California following the July 21, 1952 Kern County earthquake (M_L = 7.7). The tower system had not been designed to resist earthquake-induced lateral loads. (Photo by E. Varner. Courtesy of the National Oceanic and Atmospheric Administration, Environmental Data Service.)

$$C = \frac{0.05}{\sqrt[3]{T}} \tag{7}$$

but this was revised to Equation 3 by the SEAOC to "more realistically reflect the expected dynamic response of real structures in areas of highest seismicity,"[934] The Los Angeles Building Code[938] currently makes use of Equation 3, whereas Equation 7 is used in the San Francisco City and County Building Code.[939]

5. S or Site-Structure Resonance Factor

The S factor was introduced into the lateral force formula (Equation 2) in the 1976 edition of the UBC and is intended "to account for the variability of site conditions affecting resonance."[934] Soil amplification factors were once recognized in the Uniform Building Code, wherein the lateral force requirements were increased for "poor" foundation conditions based upon allowable design soil pressure. However, the soil amplification factor was removed from the code because of "the lack of knowledge as to what the factor or factors should be, and the lack of adequate instrumental records on varying soil types."[943] It was then left to an engineer's experience and judgment to coordinate a proposed structure's design with a particular foundation type.

That site conditions could affect ground motion has long been a recognized fact in seismology and is founded on the sporadic patterns of building damage observed after many destructive seismic events. A number of postearthquake surveys have indicated that works of construction of similar design and in close proximity suffered dramatic differences in shaking-induced damage because the structures were sited on different types of ground (e.g., alluvium versus bedrock, thick versus thin alluvium, dry versus saturated alluvium, etc.)*[502,512,516-533]

Although many aspects of ground motion and resulting building responses are not fully understood, the Seismology Committee of the SEAOC has commented about what is now accepted by one school of thought about these two parameters.[934]

* The general relationships between building damage and different site conditions for several earthquakes are summarized in the "Ground Shaking" section of Volume I, Chapter 3.

There is evidence to show that response is a function of soil depth and soil type.* Greater damage is likely to occur when the fundamental natural period of vibration of a structure is similar to that of the soil deposit on which it is constructed. Thus low, short-period buildings tend to suffer greater damage due to earthquake shaking when they are located on short-period soil deposits. Conversely, very tall multistory buildings tend to suffer greater damage when they are located on long-period deposits. In such cases, a quasi-resonance condition between a structure and the underlying soil deposit may develop, producing stronger response in the structure with increased potential for serious damage.

The procedure for accounting for local soil conditions is to vary the value of the S factor, dependent upon the degree of similarity between the period of vibration of the proposed structure (T) and the characteristic period of the site (T_s). The procedure for determing T_s is contained in Appendix E of Volume III. Two formulas are used for determining S, one being applicable when T/T_s is 1.0 or less and the second when T/T_s is greater than 1.0 (Formulas 12-4 and 12-4A, Appendix D). The value of S cannot be less than 1.0. The value of S is set at 1.5 when T_s cannot be properly established (Appendix D).

6. W or Weight Factor

The weight factor represents the effect of the inertia mass and for buildings is equal to the total dead load except in storage and warehouse occupancies where W is equal to the total dead load plus 25% of the live load. The SEAOC recommends that, for structures other than buildings, W should equal the dead load plus the live load that would be expected to be a part of the structure at the time of an earthquake. In the case of a container filled with liquid, such as a water tank, the live load would equal the weight of the liquid.[934]

7. Recent Changes in the Base Shear Equation

The I and S factors were introduced by the SEAOC in 1974/1975 and used in the 1976 edition of the Uniform Building Code. The base shear equation in several previous editions of the UBC was

$$V = ZKCW \qquad (8)$$

with the occupancy-importance and site-resonance coefficients considered as unity. Their addition to the base shear formula produces an increase in the required lateral design forces. Assume, for example, that the I and S coefficients are 1.5 each and the Z, K, and C factors remain constant for Equations 2 and 8. This would mean that Equation 2 would yield a total base shear 2.25 times larger compared with Equation 8.

B. Distribution of the Total Lateral Force

The total lateral force at the base of a structure, determined by Equation 2, is not to be distributed uniformly over the height of a structure. The distribution is determined in accordance with three distribution formulas (12-5, 12-6, 12-7; Appendix D).

C. Lateral Force on Elements of Structures — Equivalent Static Analysis

In addition to a structure being able to resist lateral forces, all elements attached to structures and their anchorages must be designed to resist lateral forces (F_p) according to:

$$F_p = ZIC_pSW_p \qquad (9)$$

* The natural period of the ground increases with increasing depth of soft alluvial fill.

FIGURE 10. *Parapet damage* to structures in Bakersfield, California following the July 21, 1952 Kern County earthquake ($M_L = 7.7$). (Courtesy of the National Oceanic and Atmospheric Administration, Environmental Data Service.)

where C_p = numerical coefficient for the horizontal force factor for elements of a structure (Table 23-J, Appendix D), W_p = weight of a portion of a structure, and Z, I, and S as defined in Equation 2.

Because the failure of building *parapets** and exterior *cornices*** poses a life-hazard to pedestrians and people in vehicles (Figure 10), these features are assigned a C_p value of 1.0. This would mean, for example, that if the Z, I, and S parameters were determined to be unity, a parapet or cornice would have to be designed and constructed to resist a constant lateral force equal to its own weight. In the case of a parapet, the direction of force is normal to its flat surface. The force can come from any horizontal direction for a cornice.

One item included in this UBC provision that is not a part of a structure is the masonry or concrete fence. This feature carries a C_p value of 0.2. If, for example, the Z, I and S factors were unity, the fence would have to be capable of resisting a lateral force normal to its flat surface equal to two tenths its weight. Past earthquakes have shown that masonry and concrete fences are extremely vulnerable to collapse if they are not adequately reinforced.

D. Minimum Earthquake Forces for Structures and Distribution of Lateral Forces — Dynamic Analysis Method

For structures having irregular shapes, large differences in stiffness between adjacent stories, or other unusual framing systems, the lateral seismic force at the base of a structure and the distribution of this force throughout its height are established by the

* Portions of side walls extending above the roof line. Some serve as false fronts while others are intended to serve as protection for firemen.[27]

** Projections and appendages usually found at or near the top of some buildings.

FIGURE 11. The 48-story (plus 65-m spire) Transamerica Pyramid in San Francisco. The establishment and distribution of lateral forces for this structural steel frame building were determined by *dynamic analysis*. The windowless protrusions rising from the 29th story house elevators on one side of the building and a stair well and smoke tower on the opposite side. (Courtesy of Dale Honeycut, Northern Arizona University.)

dynamic analysis method (Figure 11). Basically, dynamic analysis (1) estimates the seismic waveforms that are expected to be produced at the base of a structure and (2) determines the lateral resistance by calculating the response vibration of the proposed structure.[930] Dynamic analysis does not have a single solution, as does equivalent static analysis, but rather a separate solution is obtained for each of a series of time intervals during a structure's displacement history induced by the seismic loading.[927,929] A computer must be used to calculate the dynamic responses of a structure because of the sophisticated nature of this method of analysis.

As described by Dowrick[927] and Lord,[944] there are three types of dynamic analysis:

direct integration, normal mode, and *response spectra.** The direct integration of the general equations of motion is a technique applicable to coupled, uncoupled, elastic, or inelastic systems vibrating under any type of loading configuration. Normal mode analysis is used only for linear elastic systems vibrating under loads having any time function. The response spectra technique plots the maximum value of any response parameter against the vibration period for a linear, elastic, single-degree-of-freedom system (i.e., one vibration mode in a given direction). The direct integration technique provides the most complete dynamic analysis, but it is the most expensive to carry out.

The following are important steps for the dynamic analysis of a proposed structure.[945]

1. A detailed mathematical model is made of the entire structural frame that will resist the earthquake forces. The model must accurately describe the physical properties of all members and connections which make up the frame. These data are programmed into a large digital computer and stored as a detailed and exact "picture" of the stiffness, strength, and dynamic characteristics of the building.
2. Very detailed accounts of the time history of earthquake motions are programmed into the computer. This is accomplished by dividing the time record into small increments of a second with the exact motion-state existing at each of the time intervals. Both actual recorded motions from strong-motion accelerographs and simulated motions generated by computer programs are used to extend the limited knowledge of ground motions for actual earthquakes.
3. The computer solves the motion equations for the building model at each interval of motion time for every motion history. Building motion, building distortion, and building forces are determined for every part of the building at each interval of the earthquake ground motion.
4. The model is subjected to computer analysis on a repeated basis. The initial trial design is based on general dynamic knowledge. This design is subjected to selected ground motions, and its response to those motions is examined. Modifications to improve the response are made, and the revised model is subjected to the given ground motions. Several design trials are usually required to obtain an optimum design for the prescribed conditions.

Lord[944] has reported on the real-time simulation of a 52-story steel-frame building responding to various earthquake ground motions. The tower was modeled mathematically and excited by several earthquakes, and dynamic analysis obtained the time history of the structure's responses. As one example, the response for the north-south component of the May 18, 1940 Imperial Valley, California earthquake (M_L = 7.1) was simulated in real-time on a cathode ray tube, and the motion of the building (animation technique) was recorded on 16-mm movie film. The movie portrays several sequences of the structure's seismic response.

Lord[944] proposes that this type of dynamic analysis and simulation are significant to both the client and the structural engineer. Client benefits include the following:

1. Determines the degree to which a building will be resistive to ground shaking
2. Helps to determine an earthquake risk analysis for a particular structure; by this, (1) it would be possible to indicate the possibility of collapse by the maximum ground shaking that would be expected at a given site, (2) it would identify damage cost estimates for various types of ground motion, and (3) it might establish a monetary damage estimate

* The three types of dynamic analysis are described in detail by Dowrick.[927]

3. Permits a safer structure to be designed at little or no extra cost, thereby reducing liability risks and minimizing the business interruption resulting from a severe earthquake
4. Produces a structure designed to diminish the human discomfort level (e.g., from whipping motions) during a severe earthquake
5. Provides graphic assurance of a structure's stability for various earthquake-induced motions

The advantages to the structural engineer include the following:

1. Provides a check on manual sizing calculations (e.g., maximum drift coefficient over the height of the building)
2. Delineates the essential dynamic characteristics of a structure
3. Permits the engineer to "tailor" a structure to a uniform strength over its entire height
4. Identifies the degree of drift control achieved for various levels of earthquake intensity, thereby establishing clearance requirements for items such as exterior solid paneling
5. Permits the engineer to design a structure to withstand seismic forces in excess of code minima at little or no extra cost
6. Enables the engineer to obtain a visual and numerical "feel" for a structure

According to Muto,[946] one of the advances making it possible for high-rise buildings to be constructed in Japan is dynamic computer modeling. He notes that in the past it would have been foolish to risk building a high rise not knowing its dynamic responses to varying intensities of ground motion. The technique permits engineers to analyze the impact of ground motions on their proposed designs and suggests ways in which seismic-resistant design can be optimized. Los Angeles requires a dynamic analysis for all buildings higher than 48 m.[938]

E. A Sampling of Additional UBC Seismic Provisions

The UBC specifies that all portions of a structure shall be designed and constructed to act as an integral unit for resisting lateral forces unless the units are separated structurally by a sufficient distance to avoid contact under *lateral deflections* or *drift* from either seismic or wind forces. If this is not taken into account, *pounding* or *hammering damage* can result along adjoining walls. This type of damage occurs when the individual units of a building have different modes of response, and hence, different deflection characteristics. According to Hauf,[947] the condition for pounding can be crucial when low wings adjoin a high tower (Figure 12), wings of widely different masses and extent intersect, or very long buildings incorporate connection joints to accommodate movements caused by temperature changes. Pounding can also occur when the side walls of adjacent buildings are in close proximity or in contact with each other.

Unless a specific design is submitted for approval, all masonry and concrete chimneys, including those attached to single-family residences, in Seismic Risk Zones 2, 3, and 4 must be reinforced throughout their full height and anchored at each floor or ceiling line, except when constructed completely within the exterior walls of a building. Reinforcing and anchorage details are located in Chapter 37, Section 3704 of the UBC.[937] Based upon past experience in American earthquakes, the unit that often fails in a building that is otherwise undamaged is the unreinforced and improperly anchored masonry or concrete chimney (Figure 13). The collapse of chimneys represents a life-hazard for those out-of-doors when an earthquake strikes.

FIGURE 12. *Pounding* or *hammering damage* between the ballroom (left) and the 14-story Anchorage-Westward Hotel following the March 27, 1964 Alaskan earthquake (M_s = 8.5). (From Berg, G. V. and Stratta, J. L., *Anchorage and the Alaska Earthquake of March 27, 1964*, American Iron and Steel Institute, New York, 1964, 30. With permission.)

F. UBC Quality and Design Specifications

Quality and design specifications for structural materials and their connections are specified in several chapters of the Uniform Building Code (e.g., Chapter 24 — Masonry, Chapter 25 — Wood, Chapter 26 — Concrete, Chapter 27 — Steel, Chapter

FIGURE 13. Collapsed masonry chimney from a two-story building resulting from the March 8, 1937 Berkeley, California earthquake. (Photo by F. Ulrich. Courtesy of the National Oceanic and Atmospheric Administration, Environmental Data Service.)

28 — Aluminum).[937] Material specifications are based upon nationally accepted standards, and there is usually little modification for their use in regions where seismic-resistant design is required.

V. SPECIALIZED SEISMIC PROVISIONS

Because the primary function of the earthquake provisions of the Uniform Building Code is to guard against major structural failures, various enities of government have enacted special provisions to minimize damage to structures from ground shaking. In addition, special regulations have been adopted to restrict the building of certain types of structures on or near active fault zones. These ordinances are a reflection of the seismic risk perspectives of a community, and most have been enacted in California. It is beyond the scope of this book to review all specialized seismic regulations, but a sampling has been selected for discussion.

FIGURE 14. Collapsed roof and walls of Jefferson Junior High School building following the March 10, 1933 Long Beach, California earthquake (M_L = 6.3). (Photo by Captain T. J. Maher. Courtesy of the National Oceanic and Atmospheric Administration, Environmental Data Service.)

A. California Public School and Hospital Buildings

On March 10, 1933 an earthquake (M_L = 6.3) struck the Long Beach-Whittier-Compton area and destroyed or seriously damaged many buildings, including most of the school buildings (Figure 14). It was most fortunate that the earthquake struck at 5:54 p.m. on a Friday when there were only a few people in the buildings. Damage to school buildings characterized by exterior walls of brick or hollow clay tile with wood roofs and supported floors was most extreme.[948] Noting that the damage could have been much less if special design and construction practices had been followed, the California Legislature enacted the *Field Act (School House Safety Act)* approximately 1 month after the earthquake.[940,949]

In overview, the Field Act requires: (1) a California-licensed architect or structural engineer to prepare plans and supervise all phases of construction for public school buildings, including junior colleges, (2) the Division of Architecture (now the Office of Architecture and Construction), State Department of Public Works to approve or reject the plans and specifications, (3) construction to be continuously inspected by a party acceptable to the architect, structural engineer, and the state, and (4) all parties concerned with the supervision of construction to provide affidavits certifying that all construction is in conformance with the approved plans and specifications; a false statement is treated as a felony.

The Schoolhouse Section of the Office of Architecture and Construction enforces the Field Act by the following methods of operation.[949]

1. Plans, specifications, and calculations prepared by architects or structural engineers in private practice are submitted, along with the application and fee...
2. Comments are marked on the plans and specifications by the State's structural engineers for conformance to the regulations, tempered with engineering judgment, and returned for consideration.

3. Comments on the plans and specifications are discussed jointly by the designers and State personnel. A list of materials to be tested and special inspections is approved upon receipt of a copy of corrected plans and specifications. A contract may then be let by the school board.
4. An inspector, who is employed by the school board and provides continuous inspection while acting under the direction of the architect or structural engineer, is approved and construction is started. The Schoolhouse-Section-office field representative periodically visits the construction site to review for possible design and construction errors. Architects and engineers in charge of construction also periodically visit the site. Change orders and addenda proposed by the architect or structural engineer are reviewed and approved or rejected by a State representative.
5. Periodic verified reports are received from all who are concerned with the supervision of construction. On the usual project these reports are received from the architect, structural engineer, inspector, contractor, and from any special inspector and testing laboratory involved in the construction. When the final reports are received a letter is issued to the school board by the State, indicating that the Field Act provisions of the Education Code pertaining to safety of design and construction have been observed in the construction of the school building.
6. An advisory board, consisting of leaders among architects, structural engineers, mechanical engineers, and electrical engineers in private practice, provides advice to the Schoolhouse Section on all types of operational procedures.
7. Research is performed on building behavior, and evaluations are made of new building materials and techniques of construction.
8. Examinations of existing school buildings are made when requested.

The 1933 act was not retroactive, nor did it, then or now, apply to private schools or state colleges and universities. It also did not prohibit the construction of public school buildings on sites having a potential seismic hazard. The Field Act does not contain specific building regulations, but it directs the Department of General Services to adopt the detailed building regulations that are published in Title 24, California Administrative Code. Generally, the regulations are more complete than those of the Uniform Building Code.[949]

Through the years, there have been shifts in concern by state personnel regarding different components of a public school building.[936]

...During the early years of the Field Act, strong emphasis was placed on the structure and exterior appurtenances such as ornaments and parapets, while such items as lighting fixtures, and mechanical equipment were not matters of concern. Over the years and especially after the 1952 Kern County earthquakes, the performance of ceilings, lighting fixtures, bookcases, etc., were matters that concerned the engineers in the State Division of Architecture... and these items were more closely regulated. Later earthquake experience expanded the concerns into other areas of construction such as mechanical equipment.

The 1939 *Garrison Act*, as amended by the legislature in 1967, required that all public school buildings constructed prior to 1933 had to be inspected for safety by June 30, 1970. Structures found to be unsafe could not be used after June 30, 1975 unless they were repaired or replaced to meet the requirements of the Field Act. An important factor making it possible for communities to replace unsafe school buildings was voter ratification in 1972 of a constitutional amendment repealing the requirement of a two-thirds majority approval of bond issues to replace structures not meeting Field Act standards.[950]

Statutes enacted in 1967 and 1971 require geologic and engineering evaluations for (1) sites proposed for new school buildings and (2) sites for additions or alterations to existing buildings.[940,949] A portion of the 1967 statute reads as follows:[949]

The investigation shall include such geological and engineering studies as will preclude siting of a school over or within a fault, on or below a slide area, or in any other location where the geological characterisitics are such that the construction effort required to make the site safe for occupancy is economically unfeasible.

The merits of the Field Act have been demonstrated in every damaging California

TABLE 1

Damage Comparison to Earthquake Resistive and Nonresistive Masonry Public Schools of Kern County, California, West of Mohave for the July 21, 1952 Kern County Earthquake (M_L = 7.7)

	Number of schools damaged	
Extent of damage	Resistive[a]	Nonresistive[b]
None	11	1
Slight	6	6
Moderate	1	8
Severe	0	7
Collapse	0	1

[a] Built under conditions described in the 1933 Field Act.
[b] Built prior to 1933.

From Steinbrugge, K. R. and Bush, V. R., Earthquake Investigations in the Western United States 1931-1964, Publication 41-2, U.S. Government Printing Office, Washington, D.C., 1964, 240.

earthquake since its implementation. One of the first tests of its effectiveness came with the July 21, 1952 Kern County earthquake (M_L = 7.7). In the region west of Mohave, for example, many of the pre-1933 structures were seriously damaged, while the post-Field Act buildings suffered limited or no damage[27] (Table 1). Meeham[951] analyzed the effects of the February 9, 1971 San Fernando earthquake (M_L = 6.4) on school structures and came to the following conclusions:

1. There was damage to unrehabilitated pre-Field Act structures in areas having a modest amount of shaking.
2. Virtually all of several thousand post-Field Act buildings in the shaken area suffered no damage of any kind.
3. In areas where ground shaking was strong, nonstructural damage occurred in some post-Field Act buildings. Most of this centered on damaged ceilings and light fixtures.
4. Several pre-Field Act structures that had been renovated to comply with the Garrison Act were not significantly damaged.

Senate Bill (SB) 519, enacted in 1973, is aimed at providing seismic structural safety for new hospital buildings and additions to existing structures in California so that they shall remain functional for emergency purposes during and after an earthquake.[952] SB 519 requires that the plans, specifications, and calculations for structural work for a hospital building be prepared by a structural engineer. Other requirements of the bill have been described by Clark.[953]

> The act requires the California Department of Public Health . . . to observe the construction or alteration of hospital buildings and requires that geological data be reviewed by an engineering geologist. It authorizes the State to make periodic reviews of hospital operations to assure that the hospital is adequately prepared to cope with earthquake shaking. In addition, the Director of Public Health is required to appoint a Building State Board to advise and act as a board of appeals in all matters affecting seismic safety.

FIGURE 15. The Medical Treatment and Care Unit of the Olive View Community Hospital complex in Sylmar, California following the February 9, 1971 San Fernando earthquake (M_L = 6.4). Three of four stair well and recreation towers overturned and pushed into the basement. Note the damage to first-story columns. There were three casualties at this facility. One person was killed as a direct result of the earthquake, and two people under intensive care died when an earthquake-caused power failure severed the operation of all electrical life-support systems. (Courtesy of James L. Ruhle and Associates, Fullerton, Calif.)

This board shall consist of 11 members . . . from the fields of structural engineering, architecture, engineering geology, soils engineering, and hospital administration — and 6 ex officio members — Director of Public Health, State Architect, State Fire Marshall, State Geologist, Chief of the Bureau of Health Facilities Planning and Construction, and the Chief Structural Engineer of the Schoolhouse Section of the Office of Architecture and Construction; ex officio members are not entitled to vote.

The impetus for this legislation was the February 9, 1971 San Fernando earthquake (M_L = 6.4). Three earthquake-resistant designed and constructed hospitals (Olive View,[954-958] Pacoima Memorial Lutheran,[959,960] and Holy Cross[961,962]) were severely damaged in the meizoseismal area (Figure 15). At the San Fernando Veterans Administration (VA) Hospital complex, four buildings completely collapsed from ground shaking[963-965] (Figure 16). The collapsed buildings (1) had a skeleton concrete frame, concrete floors, and unreinforced hollow tile filler walls and (2) were constructed in 1925, years before earthquake-resistant regulations were established in the U.S. Because the buildings had no seismic-resistant design, they were unable to resist the horizontal ground motions. The death toll at the VA complex was 44. The total casualty count for the San Fernando earthquake was 65. In 1972, the Veterans Administration decided to abandon the site, and most of the buildings were demolished by 1975.[965] The San Fernando earthquake provided the catalyst for the VA to initiate a number of investigations to develop requirements for earthquake-resistant design criteria for all VA hospitals. This program is described by Bolt et al.[965]

FIGURE 16. Rescue operations at the San Fernando Veterans Administration Hospital complex following the February 9, 1971 San Fernando earthquake (M_L = 6.4). The collapse of four pre-1933 unreinforced masonry buildings was responsible for 44 deaths. Newer buildings and additions in the complex that were designed and constructed in accordance with earthquake-resistant provisions did not collapse. (Los Angeles City Department of Building and Safety photograph. From Brugger, W., The San Fernando, California, Earthquake of February 9, 1971, Geological Survey Professional Paper 733, U.S. Government Printing Office, Washington, D.C., 1971, 218.)

The Medical Treatment and Care Unit of the Olive View Community Hospital complex in Sylmar was designed and constructed in accordance with the Los Angeles County Building Code — 1965 Edition which had essentially the same seismic regulations as the Uniform Building Code — 1963 Edition for Seismic Risk Zone 3. The building was comprised of four symmetrical five-story wings (cross-shaped plan) with four structurally separated stair well and recreation towers at the end of each wing.[954,957] The reinforced concrete structure was built on alluvial fan deposits of unconsolidated sands and gravels.[954]

The hospital complex was located approximately 10 km southwest of the February 9 epicenter and 5 km from Pacoima Dam, where an accelerograph recorded a maximum ground acceleration of 1.25 × g for both horizontal components and a vertical acceleration of 0.7 × g[544,954] — the highest ground accelerations ever recorded. The horizontal acceleration of the ground motion probably exceeded 0.5 × g at the hospital site.[955] The ground motions greatly exceeded code determined values and produced stresses above the ultimate capacities of many structural members.[955] Structural damage included the following examples:[957]

1. The first story columns above the basement suffered severe damage (Figure 15), causing the first story to lean about 45 cm to the north. The lateral-force bracing for the basement and first story consisted of a concrete moment-resisting frame.

Damage to the upper four stories was only moderate where the bracing system consisted of reinforced concrete shear walls.*

2. Certain areas of the basement collapsed. This was more common in basement areas that extended beyond the structure, where dirt and plantings were supported by the basement roof. Some perimeter basement columns "punched through" the basement roof.
3. All towers except the north tower overturned and pushed into the basement area (Figure 15). The north tower remained standing, but leaned to the north by about 0.6m.

Degenkolb[958] describes the apparent inadequacies of the moment-resisting frame and the consequences of its inability to resist the lateral forces for this earthquake.

> Observation 3 concerns the use of frame action concrete framing (bending of beams and columns) to resist major earthquake forces. The October 1, 1969, Santa Rosa earthquake gave a preview of trouble to come when very well designed structures had large deformations causing much nonstructural damage in a comparatively small (Richter 5.6) earthquake and caused spalling in 80 percent of the columns. With the present somewhat larger (but not major or great) earthquake, the example of the Olive View . . . main building . . . should give engineers and code-writing authorities nightmares. Even though the building remained standing, all but one of the exits were unusable, and the emergency power equipment was inoperative owing to building failure. A slightly longer earthquake with the same amplitude of motion or magnitude of accelerations would probably have caused collapse of the main structure with a greater loss of life. Some more stable method of framing must be devised.

This $25 million building was damaged beyond repair and was demolished shortly after the earthquake.

B. Rehabilitation of Unreinforced Buildings — Long Beach, California

Long Beach presently has about 850 nonresidential buildings that were constructed prior to the enactment of seismic building regulations in California. Because these unreinforced concrete and masonry buildings are the most vulnerable to collapse during moderate or strong earthquakes, the Long Beach City Government added Subdivision (Chapter) 80 to the municipal building code as a mandatory means for reducing the seismic risk of these hazardous structures. As defined in the Long Beach Municipal Code,[966] the regulations of Subdivision 80 "define a systematic procedure for identifying and assessing earthquake generated hazards associated with certain existing structures within the City and to develop a flexible, yet uniform and practical procedure for correcting or reducing those hazards to tolerable hazard levels." The following discussion describes the major components of this subdivision.[966-968]

All concrete and masonry buildings constructed prior to January 9, 1934 are assigned a rating or *Hazardous Index* (i.e., relative degrees of hazard) which is used to establish a grade for determining the length of time a building can exist without corrective repairs or demolition. The grading of a building is done by personnel from the Department of Planning and Building and consists of an evaluation based upon the examination of building plans and specifications, a visual inspection, and an evaluation of the occupancy classification and occupant load. The evaluation also includes analytical procedures for determining the ability of a building's primary structural system to resist horizontal loads.

The structural analysis consists of a comparison of the lateral force-resistance capac-

* Composite type of frame (i.e., mixture of flexible and rigid frame components) described in the "Types of Structures and Structural Materials" section of this chapter.

ity of a pre-1934 building (V_{CAP}) to the required lateral force-resistance capacity of a similar type of building designed and constructed under the regulations set forth in the 1970 Uniform Building Code (V_{REQ}). The comparison is expressed in terms of a *seismic capacity ratio* (R_S):

$$R_S = \frac{V_{CAP}}{V_{REQ}} \qquad (10)$$

At least five elements of the structural system are evaluated and a critical or minimum ratio is determined. The following elements are evaluated:

$$R_{Walls} = \frac{V_{CAP}}{R_{REQ}} \quad \text{(stability of vertical walls)} \qquad (11)$$

$$R_{Anchorage} = \frac{V_{CAP}}{V_{REQ}} \quad \text{(anchorage of walls perpendicular to diaphragm)} \qquad (12)$$

$$R_{Diaphragm} = \frac{V_{CAP}}{V_{REQ}} \quad \text{(horizontal diaphragm capacity)} \qquad (13)$$

$$R_{Connections} = \frac{V_{CAP}}{V_{REQ}} \quad \text{(shear connections parallel to shear or moment resisting element)} \qquad (14)$$

$$R_V = \frac{V_{CAP}}{V_{REQ}} \quad \text{(shear or moment resisting element)} \qquad (15)$$

By using the seismic capacity ratio (R_S), the Hazardous Index (H_I) of a building is determined by the relation:

$$H_I = A(1 + \frac{200}{O.P.})R_S \qquad (16)$$

where A = occupancy classification: A = 50 for emergency buildings (e.g., fire, police, hospitals, restrained or nonambulatory occupancies, water, power, garaging of emergency vehicles, medical warehouse), A = 80 for public assembly, schools, colleges, day care centers, apartments, hotels, commercial retail buildings, food storage, industrial with hazardous contents, A = 100 for offices, garages, industrial buildings, work shops, warehouses. O.P. = occupancy potential, where an occupant load is computed based upon the building area used and occupancy Table 33A of the Long Beach Municipal Code. For buildings in Fire Zone 1 and adjacent to a public sidewalk, the occupancy potential is increased by 20%.

The grading consists of three hazardous levels and is established as follows:[968]

> Excessive Hazard — Grade I — shall consist of that approximately 10% of the buildings occupying the lowest portion of the Hazardous Index.
>
> High Hazard — Grade II — shall consist of that approximately 30% of the buildings occupying the middle portion of the Hazardous Index.

Intermediate Hazard — Grade III — shall consist of that approximately 50% of the buildings occupying the highest portion of the Hazardous Index.

Immediate Hazard will automatically place a Grade II or Grade III building into an Excessive Hazard — Grade I classification until such Immediate Hazard is removed, anchored, reconstructed, etc. Immediate Hazards are unreinforced masonry parapets, appendages, chimneys, towers, equipment, etc. adjacent to sidewalks and alleys or adjacent to smaller buildings including unreinforced masonry walls and parapets more than one story above an adjacent building.

The owners of Grade I buildings are notified to proceed with corrective repairs or to demolish the structure as soon as the rating is established. Owners of Grade II and Grade III buildings will be notified on or after January 1, 1981 and January 1, 1988, respectively, to proceed with corrective repairs or demolition.

Buildings placed in a particular hazardous grade may be changed to a lesser grade when corrective measures are completed or when there has been a modification of the use of occupancy potential. If the initial hazard assessment results in a solution "virtually equal" to that required by the Uniform Building Code, or if repairs are accomplished to comply with the UBC, a building is deemed as having no seismic hazard.

The city's hazard notification apprises the owner of the hazard grade of the building, the grievance procedure to be followed if the owner disagrees with the grading, and that the assigned grade will be recorded with the County Recorder after 60 days unless a change in grade has been initiated.

The following quotation summarizes the intent of Subdivision 80 to the owners and occupants of pre-1934 buildings.[969]

Economically, such rehabilitation and renovation is expensive. For existing hazardous structures, the cost of remedial work can amount to a relatively large percentage of total value of a structure, and the benefit-cost ratio, therefore, may be relatively small when considering property improvements for earthquake resistance. However, the social value in reduction to the threat of life loss justifies the existence of Subdivision 80. Furthermore, Subdivision 80 provides interim measures which can be instituted to reduce occupancy and use of such buildings. As a means of expediting the removal of these buildings, numerous redevelopment projects are now being proposed and considered in and around the Long Beach central business district. Removal of existing unsafe structures can best be accomplished by replacing them with new developments. In this way the safety problems can be resolved without an adverse economic impact upon the City or property owner.

With the exception of this rehabilation program in Long Beach, there is no legislature statute in California requiring nonpublic pre-Riley Act buildings to be structurally strengthened to resist earthquake forces.[970]

C. Rehabilitation of Existing Parapets and Appendages

Several jurisdictions in California, including Beverly Hills, Burbank, Glendale, Los Angeles, and San Francisco, have retroactive parapet and appendage correction ordinances. With the exception of San Francisco, the programs have been successfully completed in the above cities. The programs require removal or strengthening of all elements along building fronts or above public ways that "might break loose and fall during an earthquake."[970]

The Los Angeles program, the first of its kind, was initiated in 1947. According to Abel,[970] the Department of Building and Safety systematically surveyed all pre-1933 buildings, and when hazardous elements were discovered, building owners were served with notices to correct any unsafe conditions. At the time of the 1971 San Fernando earthquake, approximately 21,000 buildings had been surveyed and corrected, and the results were encouraging.[943] A typical parapet corrective measure is shown in Figure

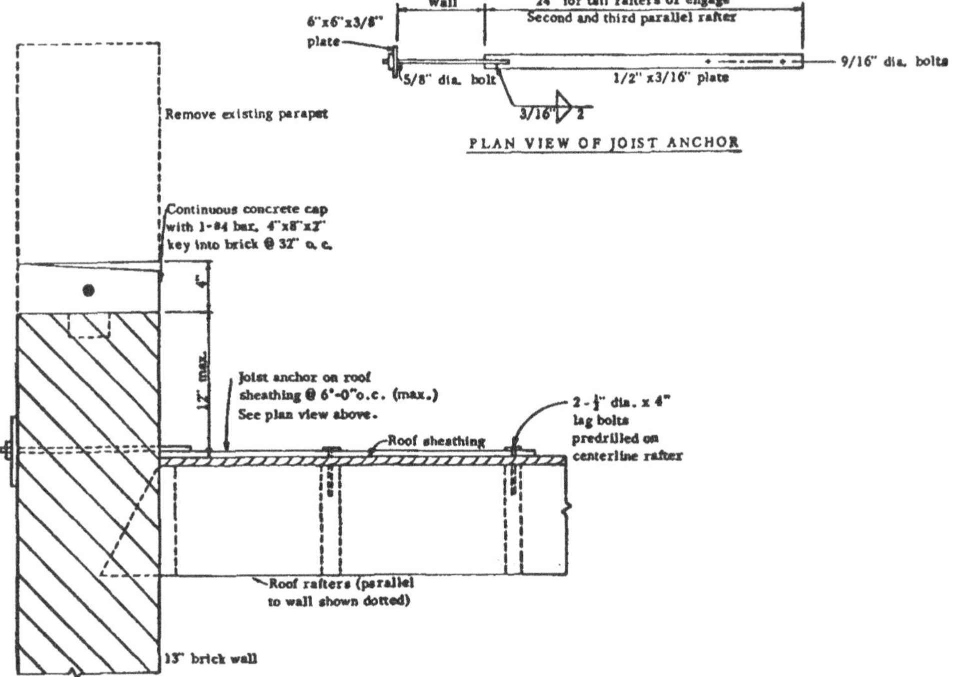

FIGURE 17. Typical *parapet corrective measure* for a pre-1933 masonry building in Los Angeles. In general, the parapet (1) height was reduced to 30.9 cm (12 in) above the roofline and (2) anchored to the roof framing by steel straps and bolts.[970] (From Abel, M. A., San Fernando, California, Earthquake of February 9, 1971, Vol. 1 (Part B), Benfer, N. A., Coffman, J. L., and Dees, L. T., Eds., U.S. Government Printing Office, Washington, D.C., 1973, 639.)

17. For the Los Angeles program, a "1-year grace period was allowed, public acceptance was good, and very little resistance by building owners was encountered."[970]

Although the San Francisco parapet ordinance was passed into law several years ago, it was not implemented until just recently because of a lack of funds for inspections.[943] Consequently, tens of thousands of potentially dangerous parapets and appendages still exist in the city.

D. Fault Easements — Portola Valley, California

The community of Portola Valley occupies a site within the San Andreas fault zone southeast of San Francisco. To protect its citizens from the faulting hazard, the Town Council enacted an ordinance that establishes special *building setback lines* for faults passing through the community. The following passages describe the provisions of this ordinance.[971]

6209.2 *Special Building Setback Lines-EF (Earthquake Fault)*. A. *Purpose*. Special Building Setback Lines-EF are established along earthquake fault traces to minimize the potential loss of property and life resulting from differential movement along such fault traces caused by tectonic forces . . .

B. *Delineation of Earthquake Fault Traces*. Earthquake fault traces are mapped as "known" locations and "inferred" locations. "Known" locations are based on surface expressions or subsurface explorations which fix the location of the trace. "Inferred" locations are based on the presence of a limited number of surface or subsurface indications of a fault trace. The actual position of the "inferred" locations is subject to wider error than the "known" location and therefore the width of potential risk band is increased.

C. *Requirements* 1) No buildings for human occupancy shall be located closer than fifty . . . feet (15.2 m) from a fault trace mapped as a "known" location.

2) Only single-family, one-story, wood-frame residences or single-family residences of different construction deemed by the Town to be of at least equivalent earthquake resistant characteristics, and buildings for other than human occupancy may be located within bands lying between fifty . . . feet and one hundred twenty five . . . feet (38.1 m) from a fault trace mapped as a "known" location.
3) When a fault trace is mapped as an "inferred" location, the setback requirements set forth in 1) and 2) above, shall be increased by fifty . . . feet (15.2 m) respectively.

D. *Measurement.* The location of a special building setback line shall be established by measurements in feet measured at right angles from the mapped fault trace as shown on the zoning map.

E. *Modification of Requirements.* When geologic studies acceptable to the Planning Commission identify an "inferred" segment of a trace at a level of accuracy equivalent to previously mapped "known" traces, such fault trace segment shall be automatically reclassified as a "known" location.

E. Seismic Regulations for Nuclear Power Plants

The U.S. Nuclear Regulatory Commission (NRC), formerly the U.S. Atomic Energy Commission (AEC), has the responsibility for evaluating the suitability of proposed sites for nuclear power plants and the suitability of a plant's design established in accordance with a proposed site's seismic and geologic characteristics. The criteria establishing the principal seismic and geologic considerations were formulated by the NRC staff and its consultants along with scientists from the U.S. Geological Survey and the National Oceanic and Atmospheric Administration. The criteria were added to Part 100 ("Reactor Siting Criteria") of Title 10 *Code of Federal Regulations* as Appendix A, entitled "Seismic and Geologic Criteria for Nuclear Power Plants," and went into effect on December 13, 1973.[972,973] The NRC summarized the intent of Appendix A in its *1973 Annual Report to Congress.*[972]

The criteria reflect advances in the state-of-the-art geologic investigations achieved since late 1971 by giving more credit to three-dimensional investigations, such as those obtained from offshore geologic surveys, in determining the extent of the zone requiring detailed faulting investigations.

The criteria describe the investigations required to obtain the geologic and seismic data necessary to determine site suitability and to provide reasonable assurance that the proposed nuclear powerplant can be constructed and operated at a proposed site without undue risk to the health and safety of the public.

Information obtained from the investigations will be used to determine the design requirements for withstanding earthquake-produced ground motion and seismically-induced floods and water waves. This information also will be used to determine whether, and to what extent, the nuclear powerplant needs to be designed for surface faulting.

The "Seismic and Geologic Criteria for Nuclear Power Plants" section of the *Code of Federal Regulations*[973] is presented in Appendix F of Volume III.

To date, there have been several siting problems involving faults, and most of these have been encountered along the California coast. The first case came to light in 1964 with the Pacific Gas and Electric Company's proposed plant at Bodega Bay, north of San Francisco (Figure 18). This facility was already under construction when it was discovered that the site was traversed by a fault of small displacement. The site is just outside the well-marked San Andreas fault zone. Construction was halted and the site forever abandoned because the NRC concluded that "there was uncertainty associated with the effects of a major earthquake involving substantial shear movement of the foundation rock at the proposed site"[972] (Figure 18).

In 1973, the Pacific Gas and Electric Company withdrew its application to the NRC to construct a two-unit nuclear power facility at Point Arena in Mendocino County because the possibility existed that offshore and onshore faulting could affect the site sometime in the future.[972,974] The U.S. Geological Survey, a consulting agency to the NRC, concluded that "even given the most careful execution of the exploration program as outlined and the most favorable return of data for efforts expended, there

FIGURE 18. Initial construction phase of Pacific Gas and Electric Company's nuclear power plant at Bodega Bay, California. The site was abandoned in 1964 because of a potential fault hazard. (Courtesy of James L. Ruhle and Associates, Fullerton, Calif.)

would remain certain areas of inadequate coverage and certain residual indeterminacies which would preclude final evaluation of the site with the degree of conservative assurance normally required for such applications.''[972]

Licensing of the Pacific Gas and Electric Company's two-unit Diablo Canyon facility, near San Luis Obispo, must await new seismic hazard reviews because a major fault was discovered a few kilometers offshore. These units are in the final stages of construction, and according to Carter,[975] the NRC might require a substantial strengthening of the two units.

Because of the ''serious and often unexpected earthquake problems'' associated with coastal sites in California, Carter[975] reports that most or all new nuclear power plants will likely be constructed in the Central Valley and Mojave Desert ''where the earthquake hazards can be more easily assessed and avoided.'' However, these are areas where already scarce supplies of fresh water would have to be used for reactor cooling purposes.[975]

In addition to siting problems in California, evidence of faulting has been discovered at the Virginia Electric and Power Company's North Anna, Louisa County site and at the South Carolina Electric and Gas Company's Broad River site.[972]

Chapter 4

BUILDING AND LIFELINE RESPONSES TO EARTHQUAKES

I. INTRODUCTION

This chapter is concerned with the *structural* and *architectural* or *nonstructural* performance of various types of building systems to earthquake-induced ground motions and the performance of *lifelines* to the faulting and ground vibration hazards of an earthquake.* Although many people equate earthquake damage solely with buildings, lifelines, which include public utility and transportation systems, can also fail with serious consequences to the public health and welfare of the stricken area. Two classes of dynamic tests that are used to determine the response of existing structures and foundation materials to exciting forces are also described in this chapter.

II. BUILDING RESPONSES TO GROUND MOTION

In overview, several noninstrumental aspects of building responses to ground vibrations have been revealed by post-earthquake engineering surveys.

- The quality of earthquake-resistant design and construction will almost entirely determine whether or not strong ground shaking will produce a disaster in terms of loss of life, economic losses, or public hardship. As stated by the Committee on the Alaska Earthquake,[976] "there is no substitute for good earthquake engineering." In addition, the likelihood for a major earthquake disaster decreases the longer earthquake engineering precautions have been in effect.
- On a case-by-case basis, structures built in conformance with earlier seismic codes may not perform as well as those structures built under the latest regulations because of code improvements and because of possible material deterioration in an older code-designed structure. However, poor workmanship, substandard building materials, and ineffective or nonexistent inspections by an enforcement agency can destroy the effectiveness of even the best designed structures for resisting lateral loads.
- For code-designed buildings, those that have a simple plan and elevation configuration, equal stiffness throughout, and sound structural connections have a more favorable response to earthquake forces than structures not incorporating these three design principles.
- Regarding nonearthquake-resistant building types and their susceptibility to damage, low-rise wood frame structures offer the best resistance against ground shaking, and buildings with unreinforced masonry construction of brick, stone, adobe, hollow concrete block, and hollow clay tile offer the least resistance (Table 1). According to Richter,[27] when the material is not unusually weak, the failure of masonry is due to imperfections in the mortar, workmanship, or design, including the omission of reinforcement. In reference to mortar without reinforcement, the mortar may have effectively held the masonry elements apart for years, but it may suddenly be required to hold the elements together during an earthquake.[977] The mortar is unable to accomplish the latter during strong shak-

* Refer to Volume I, Chapter 3 for discussions centering on the response of buildings to the faulting hazard.

TABLE 1

Hazard Comparison of Nonearthquake-Resistive Buildings[a]

Simplified description of structural types	Relative damageability (in order of increasing susceptibility to damage)
Small wood-frame structures, i.e., dwellings not over 255 m^2 and not over 3 stories	1.0
Single or multistory steel-frame buildings with concrete exterior walls, concrete floors, and concrete roof; moderate wall openings	1.5
Single or multistory reinforced concrete buildings with concrete exterior walls, concrete floors, and concrete roof; moderate wall openings	2.0
Large area wood-frame buildings and other wood-frame buildings	3.0—4.0
Single or multistory steel-frame buildings with unreinforced masonry exterior wall panels; concrete floors and concrete roof	4.0
Single or multistory reinforced concrete-frame buildings with unreinforced masonry exterior wall panels, concrete floors, and concrete roof	5.0
Reinforced concrete bearing walls with supported floors and roof of any material (usually wood)	5.0
Buildings with unreinforced brick masonry having sand-lime mortar and with supported floors and roof of any material (usually wood)	7.0 and up
Bearing walls of unreinforced adobe, unreinforced hollow concrete block, or unreinforced hollow clay tile	Collapse hazards in moderate shocks

[a] Table is not complete. Additional considerations would include parapets, building interiors, utilities, building orientation, and frequency response.

From Smith, K. R., in The Seismic Safety Study for the General Plan, California Council of Intergovernmental Relations, Sacramento, 1973, 161.

ing. Botsai et al.[933] demonstrate the characteristics of masonry construction by describing the effects of horizontal loading on a stack of bricks.

The opposite situation (regarding flexibility) is represented by a stack of unreinforced bricks whose movements result in permanent displacement of each brick when a horizontal force is applied. The stack is quickly toppled. If the bricks were cemented together with epoxy, or heavily reinforced and tied to the base so as to act as a single mass of bricks rather than as single bricks, then the stack would be very rigid and would resist the displacement forces until the mass fractured.

Because the failure of unreinforced masonry buildings has been repeatedly responsible for a large part of the loss of life and property in earthquakes, this type of construction has been called the *dead hand of tradition.*[27] The devastating effect of ground motion on unreinforced adobe construction is depicted in Figures 1 and 2.

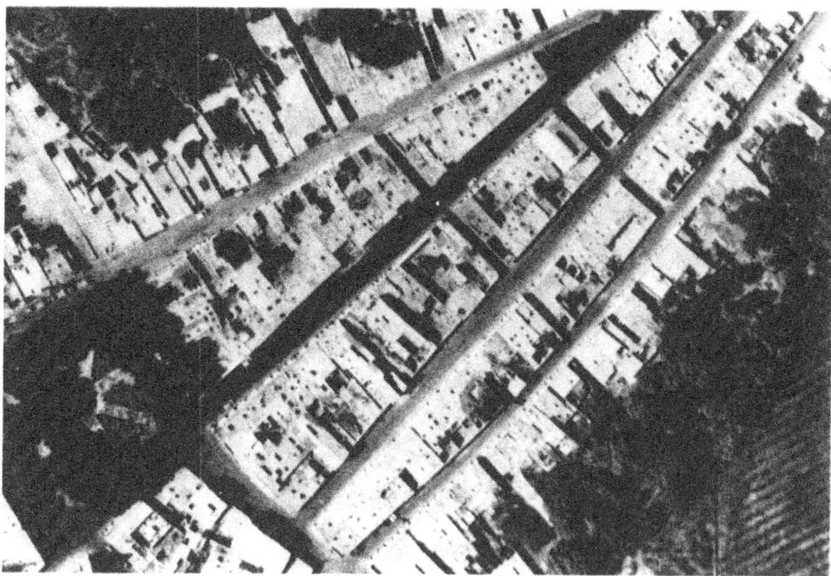

FIGURE 1. Vertical aerial photograph showing a portion of Casma before the May 31, 1970 Peru earthquake (M_s = 7.7). (From Cluff, L. S., *Bull. Seismol. Soc. Am.*, 61, 529, 1971. With permission; photo courtesty of Lloyd S. Cluff, Woodward-Clyde Consultants, San Francisco.)

Small wood-frame structures have a natural built-in resistance to horizontal loads mainly because of their high strength-to-weight ratio.[927] According to Dowrick,[927] the most common causes of the inadequate performance of wood construction to ground motions include (a) lack of integrity of the substructure, (b) asymmetrical structure form, (c) inadequate structural connections, (d) use of heavy roofs without increasing the strength of the supportive frame, and (e) deterioration in the strength of the wood due to rotting and pest attack (Figure 3).

- Shaking damage to a structure is *cumulative* if adequate repairs are not made. A building may be seriously damaged or even collapse in a moderate earthquake because it was progressively weakened by previous episodes of ground shaking. The cumulative effect can also occur with a main shock and weaker aftershocks. Steinbrugge and Bush[948] note that the "cumulative effect is not always well understood by the public who, in time, may consider a slightly loosened brick bearing wall to be safe since it survived without collapsing."
- More damage can be expected when a particular structure and the underlying foundation approach the same vibrational period. For example, tall buildings have a long predominant vibrational period (2 or more sec) and are subject to greater damage if sited on a foundation material with a long predominant vibration period, such as a thick layer of alluvium. One- or two-story buildings have shorter periods of vibration and usually suffer the greatest damage when sited on firm ground or bedrock. A structure can experience more vibration-induced damage if it rests on two types of ground as opposed to a single type.
- Short-period ground motions attenuate more rapidly with increasing distance from the epicenter than do long-period ground motions. Consequently, low-rise buildings may be damaged in the near-field and high-rise buildings at relatively large distances from the epicenter of the same earthquake (Figure 4). During the March 27, 1964 Alaskan earthquake (M_s = 8.5), tall structures in Anchorage,

FIGURE 2. Vertical aerial photograph showing a portion of Casma following the May 31, 1970 Peru earthquake (M_s = 7.7). According to Cluff,[935] Casma, located about 61 km from the epicenter, was almost completely destroyed as a result of the strong ground shaking. Nearly all of the buildings were of adobe construction. (From Cluff, L.S., *Bull. Seismol. Soc. Am.*, 61, 529, 1971. With permission; photo courtesy of Lloyd S. Cluff, Woodward-Clyde Consultants, San Francisco.)

located 130 km from the epicenter, suffered significant vibration damage,[933] whereas small-sized buildings generally escaped unscathed.[978]

- The ability of a structural system to resist horizontal loading is very dependent upon the cumulative number of stress or displacement cycles that are induced to the system; as the number of cycles increases, there is a greater need for the absorption of energy. The duration of ground motion determines the number of displacement cycles, and this motion can last from only a few seconds to several minutes. The duration of the strong component of ground motion (e.g., 0.05 × g acceleration or greater) decreases with increasing epicentral distances and varies directly with magnitude at a given distance.[933]
- Damage to nonstructural elements during lateral movements can create serious safety hazards to occupants or passersby, and the costs to repair architectural damage can exceed the costs for repairing a building's structural damage. Safety from architectural failures is usually achieved by securing the elements to the frame or floors.[934] Until recently, building design philosophy was directed primarily to the structural frame, but according to Ayers,[979] the architectural damage incurred during the 1971 San Fernando earthquake once again demonstrated the need for a refreshing review of architectural design philosophy.

> We must modify our design philosophy that a building is safe if it survives an earthquake without damage to the structural system. The structural frame may absorb the earthquake forces without failures, but the movement of the building induces significant secondary damage to nonstructural elements.
>
> A building is not safe if, during an earthquake, light fixtures and ceilings fall, elevators do not operate, emergency generators to not come on, loose objects block exits, and broken glass falls into the street. A building is not properly designed if an owner sustains huge losses due to nonstructural damage. The lessons learned by detailed studies of damage sustained by these earthquake-tested buildings must be carefully documented and widely disseminated.

FIGURE 3. Wood-frame house severed from its foundation as a result of March 10, 1933 Long Beach, California earthquake (M_L = 6.3). The studding had been weakened by termites. (Photo by Mr. Merritt; courtesy of National Oceanic and Atmospheric Administration, Environmental Data Service.)

III. POST-EARTHQUAKE DAMAGE SURVEYS

Detailed engineering damage surveys have been completed for a number of contemporary earthquakes. Five earthquakes have been selected for discussion to illustrate the response characteristics of different types of structures to ground motions.

A. July 21, 1952 Kern County, California Earthquake

The July 21, 1952 Kern County, California earthquake (M_L = 7.7) represents the largest seismic event to strike the state since 1906 and the largest in southern California since 1857; it was felt over an area of 414,400 km^2.[980] There were 12 deaths and $60 million in property damage.[981] The communities of Tehachapi and Arvin were especially hard hit. Several hundred aftershocks were recorded during the ensuing months; more than 20 had a local magnitude of 5.0 or larger and 3 were over 6.0.[980,981] On August 22, an aftershock (M_L = 5.8) struck near Bakersfield causing two additional deaths and extensive damage to many structures already substantially weakened by the main shock of July 21.[948]

The Kern County earthquake represented the first seismic event where a significant number of earthquake-resistant buildings were subjected to strong ground motions in the U.S. The following damage evaluations are for an area in Kern County bounded by Tehachapi, Bakersfield, and Grapevine.[948,982]

Wood-frame buildings were commonly used for residences, and only rarely did they suffer more damage than cracked plaster and destroyed unreinforced brick chimneys. The buildings that were seriously damaged had structural deficiencies that included no anchorages between the frame and foundation, no lateral force bracing elements, and decayed studs between the foundation and the first floor (i.e., cripple studs).

All steel structures, such as gasoline service stations, had no or only negligible damage. These buildings were highly resistant to horizontal loading because of their small

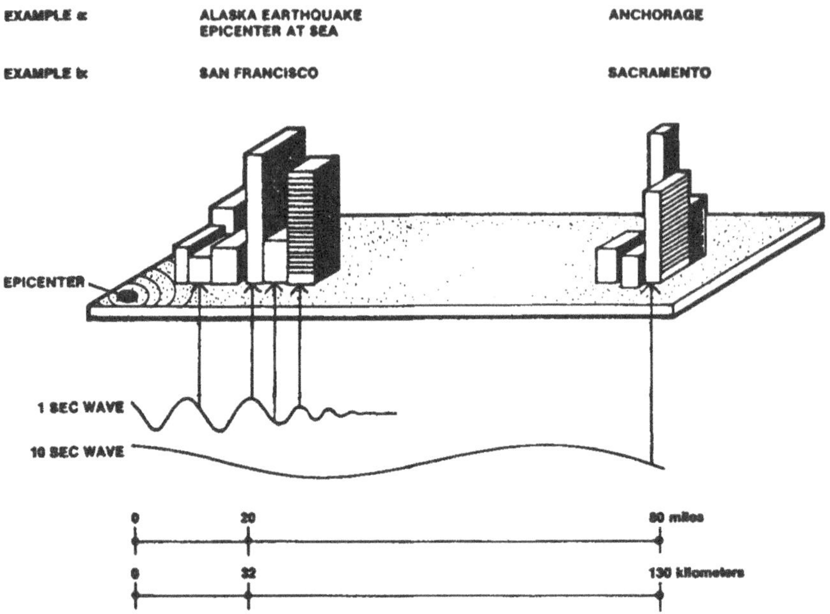

FIGURE 4. Relative wave motion effects at different epicentral distances. Note that the shorter period waves tend to die out more rapidly with distance than do longer period waves. Shorter period waves oscillate in the same frequency range as lower buildings, affecting such structures close to the epicenter. Longer period waves, which oscillate in the same frequency range as taller buildings, travel farther and can affect such buildings at relatively great distances from the epicenter. (From Botsai, E. E., et al., *Architects and Earthquakes*, AIA Research Corp., Washington, D.C., 1977, 23. With permission.)

weight and the bracing required to resist wind forces. Multistory steel-frame buildings were not common in the county, but the five-story steel-frame Haberfelde Building in Bakersfield sustained considerable nonstructural damage, especially to partitions and exterior facing. Approximately half of the monetary loss was related to pounding damage that occurred between the two units of the building.

Numerous one-story reinforced concrete buildings were found throughout the county. Those buildings having specifically designed lateral force bracing systems had, at most, minor or negligible damage. The performance of structures without bracing was reasonably good because most "were inherently strong due to small wall openings and small roof areas."[948] Damage was usually caused by inadequate ties between the roof and walls and the presence of foreign matter along construction joints* in the concrete walls. An exception was the collapse of the Cummings Valley School (Figure 5). As noted by Richter,[27] the concrete used in this building was weak and the reinforcement bars did not lap at construction joints; consequently, the structure separated into a number of individual blocks during the earthquake and collapsed.[27]

Unreinforced brick buildings with sand-lime mortar were common in the older areas of Tehachapi, Bakersfield, and Arvin. This class of construction fared poorly in the earthquake because of its inability to resist lateral forces; damage was especially severe in Tehachapi (Figures 6 and 7). By contrast, brick structures built by a technique known as *reinforced grouted-brick masonry* responded exceptionally well. Steinbrugge and Bush[948] describe the components of this construction method and the earthquake performance of one reinforced grouted-brick masonry complex.

* Junctures between poured sections.

FIGURE 5. Collapsed Cummings Valley School following the July 21, 1952 Kern County, California earthquake (M_L = 7.7). The school, built about 1910,[948] was constructed of reinforced concrete walls and wood roof. (Courtesy of National Oceanic and Atmospheric Administration, Environmental Data Service.)

. . . The building technique involves two wythes of brick laid in cement mortar. The wythes are separated by several inches and the space is filled with small aggregate reinforced concrete (called grout).

The best known example of reinforced grouted-brick masonry is the Arvin High School in Arvin. The school consisted of about 15 buildings constructed in the period 1949—1951...Reinforced grouted-brick masonry was used as the principal wall material on most major buildings. The design and construction were carried out under the requirements of California's Field Act. Minor or negligible damage was found in most buildings, and none of it constituted a life hazard. However, the 2-story Administration Building had significant damage to one 8-½-inch (21.6 cm) thick, reinforced grouted-brick wall as a result of the July 21 earthquake. Subsequent shocks increased the damage. While seriously damaged, collapse was not imminent*...The overall damage to all buildings was less than 1% of the value.

Steinbrugge and Bush[948] describe a contrasting pattern of damage in Los Angeles and Long Beach for the Kern County earthquake.

. . . Damage in Los Angeles...was generally confined to steel and concrete frame fire-resistive structures over 5 or 6 stories high. A few isolated instances of minor damage to 1- and 2-story buildings were noted, but they were not significant. This pattern of damage was opposite to that experienced in Kern County on July 21 and Bakersfield on August 22, 1952, in that there 1- and 2-story brick bearing-wall buildings were much more affected than the multistory steel and concrete frame buildings.

The explanation for this difference is that the ground motion in the Los Angeles area was generally of longer periods, which adversely affect taller buildings with corresponding longer natural periods. In other words, the motion some 70—80 miles (113—129 km) from the epicenter was such as to excite vibrations of crack-producing magnitudes in tall structures while not affecting the lower, more rigid buildings. A contributing factor was the previous damage to these tall buildings in past shocks, particularly the Long Beach shock of 1933, since effective repairs had generally never been made. No cases of structural damage were noted, and principal damage was to partitions, masonry filler walls, ceilings, marble trim, veneer, and exterior facing. It should be added that the buildings under discussion are the older ones without adequate earthquake bracing . . .

Behavior of tall buildings in Long Beach was similar to that in Los Angeles. However, it is disquieting to note rather extensive nonstructural damage to major structures, in some cases, when one considers that the

* Workmanship errors (e.g., grout core was not completely filled and the grout did not bond to the brick or reinforcement bars) were responsible for the damage incurred by this wall.[948]

FIGURE 6. Unreinforced brick Lodge Hall in Tehachapi following the July 21, 1952 Kern County, California earthquake (M_L = 7.7). Wood partition at left shows original roof height of this two-story structure. The roof collapsed after supporting walls failed. (Courtesy of National Oceanic and Atmospheric Adminstration, Environmental Data Service.)

structures were located some 100 miles (161.3 km) south of the epicenter. In the 1933 Long Beach shock these buildings, in general, suffered more extensive damage than those in Los Angeles, and the methods of repair were often equally ineffective.

B. March 27, 1964 Alaskan Earthquake

The March 27, 1964 Alaskan earthquake (M_S = 8.5) is the largest seismic event to date to test many modern types of earthquake-resistant construction. Direct vibratory damage was significant in Anchorage with minor vibratory damage widespread throughout the area of strong ground shaking (Table 7 in Volume I, Chapter 3). Anchorage, located approximately 130 km west-northwest from the epicenter (Figure 13 in Volume I, Chapter 3), bore the brunt of property damage because of its size;* municipal and private property losses amounted to $86 million,[983] 27.7% of the total earthquake losses in the state. Property damage was caused by seismic vibrations, ground cracks, and landslides (Table 7 in Volume I, Chapter 3) with the latter hazard causing the most damage[550] (see Section III.A.1 in Volume I, Chapter 3). The earthquake was responsible for nine deaths in Anchorage. According to Steinbrugge,[984] one of the important reasons for the low death figure was that earthquake-resistant design and construction techniques had been used for "practically all of the substantial buildings."

The duration of strong or damaging ground motion was estimated to have been approximately 3 min in Anchorage. This compares to estimates of 40 to 60 sec of severe shaking in the April 18, 1906 San Francisco earthquake (M_S = 8.3).**[983] Steinbrugge et al.[983] describe the impact of the extended duration of ground shaking to building damage in Anchorage.

> . . . the longer the earthquake's duration of damaging intensity, the greater will be the damage. Many repeated excursions into the yield range eventually brings destruction to steel. Hairline shear cracks in rein-

* Anchorage is the largest city in the state; at the time of the earthquake, the municipal population was 50,000 while the population for metropolitan Anchorage approached 100,000.[983]

** Durations of damaging intensity for the Anchorage and San Francisco earthquakes can only be estimates owing to the absence of strong-motion seismograph records.

FIGURE 7. Collapsed unreinforced brick bearing wall of the Juanita Hotel in Tehachapi following the July 21, 1952 Kern County, California earthquake (M_L = 7.7). Note support given by the nonstructural wood partitions after collapse of exterior bearing walls. (University of California, Berkeley, photograph; courtesy of National Oceanic and Atmospheric Administration, Environmental Data Service.)

forced concrete become larger with an extended duration and finally can bring failure. The observer's accounts of building damage...state that much of the vibrational damage and collapse occurred in the latter stages of the earthquake, showing that a duration of 1 minute or less, such as in San Francisco in 1906, might not have caused many of the collapses . . .

The predominant periods of horizontal ground motion in Anchorage were probably on the order of 0.5 sec and longer[983] (Figure 4). This would mean that the long-period motions should subject the taller buildings to larger lateral forces and significantly more vibration-induced damage than small, rigid structures. The ground-period thesis was borne out by an extensive damage survey of the city. As described by Berg and Stratta:[978] "Most of the severely damaged structures were either tall or massive. Houses and small buildings generally escaped significant vibration damage." The following descriptions are for several earthquake-engineered structures that were significantly damaged or destroyed by direct seismic vibrations.

- The J.C. Penney building was constructed in 1962 under the Zone 3 earthquake provisions of the Uniform Building Code.[983] It was a five-story reinforced concrete structure, nearly square in plan, with shear walls on three sides. The north and east walls were covered with 10-cm-thick precast reinforced concrete panels from the second floor to the roof. No lateral bracing system was used in the north wall above the first story. The floors were 25-cm-thick reinforced concrete flat plates supported on 129 cm² reinforced concrete columns.[550,978]

  Earthquake-induced torsion was the triggering mechanism for the castatrophic failure of the building (Figure 8). Torsional forces were not initiated in the first story because the shear wall bracing system was constructed along all sides. However, because the north wall was structurally open above the first floor, the center of rigidity was near the south wall, far from the center of mass. Consequently, large torsional forces were produced when the U-shaped shear wall system was subjected to east-west horizontal loading.[978,983] Examples of damage were (1) the west wall failed at the second floor, (2) most of the precast panels were shaken loose, dropping to the streets, (3) the floor slabs along the north elevation were sheared at their connections to one column at all floor levels above the second,

FIGURE 8. Northeast corner of the five-story reinforced concrete J.C. Penney building in Anchorage following the March 27, 1964 Alaskan earthquake (M_s = 8.5). Most of the rubble has been cleared from the streets. (Courtesy of George Plafker, U.S. Geological Survey.)

and (4) the northeast corner of the building collapsed[978] (Figure 8). The building was torn down shortly after the earthquake.

The following passage is an eyewitness account of the response of the J.C. Penney building to the 1964 earthquake:[978]

> Mr. E. T. Dimock was on the first floor of the undamaged Hoblit Building across the street to the east of the Penney Building at the time of the earthquake. He first heard the earthquake sounds, then felt the shaking, and then walked about 20 feet (6.1 m) to get out of the building. Thereafter, he was standing on the sidewalk across the street from the Penney Building where he could observe the east and south elevations. As the earthquake motions appeared to become more intense, he first noted that the Penney Store started to "twist", with movement along the second floor construction joints...This twisting continued for the duration of the earthquake. As he recalls, the first precast panel to fall was at the north end of the east wall, with this occurring near the middle of the earthquake. Then the panels over the east wall doorway fell, hitting the canopy, and ricocheting into the street.

- The three-story First Federal Savings and Loan building was constructed in the early 1960s and measures 15.2 by 39.6 m in plan. The frame is structural steel, and the roof and floors are reinforced concrete slabs supported by wide-flange steel beams.[978,983] To resist horizontal forces in the narrow direction (east-west), a reinforced concrete block wall on the north elevation and a brick panel with steel X-bracing (shown in Figure 9) on the south wall were used. Resistance to horizontal forces in the long direction (north-south) was to have been provided by a reinforced concrete brick wall on the west wall and two brick panels on the east wall.[978]

  A postearthquake view of a portion of the building is shown in Figure 9. The types of damage included (1) sheared masonry and brick walls, (2) deformed anchor bolts in the steel frame, and (3) one weld failure in an X-bracing connection. There was only a limited amount of glass breakage, even though the east and south elevations were mainly glass (Figure 9).[550,978]
- The control tower at Anchorage International Airport was a six-story reinforced concrete structure attached to a two-story terminal building. The external walls were covered with metal panels which provided no significant lateral force brac-

FIGURE 9. Three-story steel-frame First Federal Savings and Loan building in Anchorage following the March 27, 1964 Alaskan earthquake (M_S = 8.5). Demolition of masonry exterior walls has been completed enabling repairs to be made. Note the X-bracing in the south wall and the limited amount of glass breakage on the south elevation. (From Berg, G. V. and Stratta, J. L., *Anchorage and the Alaska Earthquake of March 27, 1964*, American Iron and Steel Institute, New York, 1964, 36. With permission.)

ing. The tower had interior concrete block walls that were nonstructural. Hence, the system for resisting lateral loading was entirely in the reinforced concrete frame. The ground motions caused the tower to completely collapse (Figure 10), killing one person and injuring a second.[978,983] With this type of structural system, the total lateral load had to be borne by the reinforced concrete frame; the unreinforced interior walls could not act as a second line of defense. Consequently, once the external bearing walls failed the building collapsed.

According to Berg and Stratta,[978] there is a possibility that the tower suffered structural damage as a result of the October 3, 1954 earthquake (M_L = 6.75), located approximately 80 km south of the city. If so, the strength of the building would have been impaired since that date.[978]

• One of the most spectacular examples of total collapse was associated with the Four Seasons apartment building (Figures 11 and 12). This structure was built in accordance with the Zone 3 earthquake regulations of the 1963 Uniform Building Code. It was a six-story lift-slab reinforced concrete building with two poured-in-place reinforced concrete cores that were designed to act as shear walls to stabilize the building against lateral forces. Stair wells and an elevator were housed in the cores.[978,985]

According to Berg and Stratta,[978] the two cores were unable to resist the horizontal forces, and they fractured in the first story. More specifically, George et al.[986] state that the cores failed because of an inadequate overlapping of reinforcing bars at their bases, allowing them to rock, and thereby severing the core and floor slab connections. Ayers et al.[987] pointedly describe the end result.

> . . . The only recognizable elements remaining after the earthquake were stair shafts and the elevator hoistings . . . it is obvious that the only place where anyone could possibly have survived would have been in the elevator car.

FIGURE 10. Collapsed six-story reinforced concrete control tower at Anchorage International Airport following the March 27, 1964 Alaskan earthquake (M_s = 8.5). (Courtesy of George Plafker, U.S. Geological Survey.)

Fortunately, the building was not occupied at the time of the earthquake. It was to have opened in mid-April 1964 as "the luxury apartment house of Anchorage."[978]

The following passage is an eyewitness account with comments by Steinbrugge et al.[983] of the collapse of the Four Seasons building:

> . . . The eyewitness account . . . by Mr. Bob Smith, who was in a one-story wood frame structure about 100 feet (30.5 m) from the building is worth repeating. He was in his front office when the earthquake occurred. He immediately noticed the walls of the Four Seasons vibrating. He then went . . . to an adjoining room to shut off a hot plate. He returned to the front office and again watched the Four Seasons vibrating. A companion stated that he thought that the apartment house would fall down if the shaking did not stop. By this time, the drawers of the filing cabinet in Mr. Smith's office had come out and had made him feel quite uneasy, so he left the structure and went outside to watch Four Seasons. Shortly before the collapse, it appeared to him that the walls were oscillating excessively. However, Mr. Smith definitely states that nothing had fallen prior to collapse...although he strongly recalls "cracks" in the walls indicative of his being able to see into the building or through it. Finally, it collapsed. As well as Mr. Smith could recall, all floors fell at one time. He was quite sure that one floor did not fall on the other, then causing the one below to fall, and so on...The ground shaking stopped almost immediately after the collapse according to Mr. Smith. It is interesting to note that all during the earthquake he had no great difficulty in standing or in walking about.
>
> The Smith account . . . indicates that the structural collapse came at or near the end of a long-duration earthquake. The duration of damaging intensity may have been as long as 3 minutes. It is easily conceivable that the structure experienced a great many cycles of motion, and any minor concrete cracks which formed during the first few cycles probably would become major cracks during this long-duration motion.

According to Berg and Stratta,[978,988] several lessons were learned from the 1964 Alaskan earthquake.

1. The inertia force originating in a structure is equal to the product of mass and acceleration. Therefore, it is advantageous to avoid unnecessary dead loads, such as a heavy roof, in order to diminish inertia forces.
2. The most rigid element in the structural frame will receive most of the lateral forces. If the element is incapable of resisting the loads, it will fail. In the First Federal Savings and Loan building (Figure 9), the most rigid element was the masonry walls. However, they failed because they were not strong enough, and the lateral forces had to be resisted by the more flexibe steel frame.

FIGURE 11. Six-story reinforced concrete Four Seasons apartment building in Anchorage before the March 27, 1964 Alaskan earthquake (M_s = 8.5). (From Berg, G. V. and Stratta, J. L., *Anchorage and the Alaska Earthquake of March 27, 1964,* American Iron and Steel Institute, New York, 1964, 20. With permission.)

3. A structure must be able to transmit seismic forces from the point of origin all the way to the underlying foundation material. A single zone of weakness along this path can be a location for failure. In the J.C. Penney building (Figure 8), once the torsional forces were generated there was no adequate means for transmitting these forces to the foundation.
4. Adjacent sections of a structure with different dynamic properties will oscillate at different periods. If too close or not adequately tied together, the segments will oscillate out of phase and hammer or pound against each other. Hammering damage of the Anchorage-Westward Hotel is shown in Chapter 3, Figure 12.
5. Connection details warrant careful attention. In order to take advantage of the energy absorbing capacity of structural members, connections should be designed so that if structural failure occurs, the failure will be in a structural member and not a connection. For example, if welds are to be effective they must be of sufficient length to distribute the transmitted force.
6. Good seismic design requires providing structural continuity or a second line of seismic resistance. For example, all lateral force resistance was confined to the two reinforced concrete cores in the Four Seasons building. When these failed, there was literally no other element to absorb the inertia forces and the building collapsed (Figure 12). By contrast, the structural steel frame in the First Federal Savings and Loan Building was capable of resisting the inertia forces after the masonry walls were sheared (Figure 9).
7. Proper design, supervision, and inspection are all important elements in any attempt to insure competent earthquake-resistant construction.

C. February 9, 1971 San Fernando, California Earthquake

The San Fernando, California earthquake (M_L = 6.4) struck on the fringe of the Los Angeles metropolitan region (over 8 million people) at 6:01 a.m. local time on February 9, 1971 (Figure 13). Its epicenter, in the San Gabriel Mountains, was just north of the San Fernando Valley (Figure 5 in Volume I, Chapter 3) and approximately

FIGURE 12. Collapsed Four Seasons apartment building in Anchorage following the March 27, 1964 Alaskan earthquake (M_S = 8.5). An adjacent two-story wood-frame building was undamaged.[947] (From Berg, G. V. and Stratta, J. L., *Anchorage and the Alaska Earthquake of March 27, 1964*, American Iron and Steel Institute, New York, 1964, 20. With permission.)

40 km from downtown Los Angeles (Figure 13). The earthquake lasted for approximately 1 min with the strong motion ($> 0.2 \times g$) persisting for 10 to 12 sec. During this brief period, 65 persons were killed, more than 2500 persons were injured, more than 24,500 structures were damaged, and the greater Los Angeles area suffered property damage estimated at $553 million[989,990] (Table 2). The timing of the earthquake was most fortunate, for if it had occurred 1 or 2 hr later, deaths and injuries would have been much greater because the freeways would have been crowded and many persons would have been in or near buildings that experienced partial or total collapse.[991]

Although the earthquake was only of moderate size, the damage to buildings and lifelines was severe in the northern part of the San Fernando Valley and in the Newhall area because the energy release was at an unusually shallow depth[992] (Figure 13). No seismographs were close enough to the epicenter to accurately determine the focal depth, but 12 km is considered reasonable.[993] The heavily shaken area was approximately 750 km^2 in size; this area contained more than 302,000 dwellings and a population of approximately 1.3 million.[994] The intensity of shaking in the meizoseismal region might well have been as severe as would be expected for an earthquake of magnitude 8.0 or larger; however, a much larger area would be affected by strong shaking for a longer period of time in a great earthquake.[992] A maximum intensity of VIII-XI was assigned to a small area in the foothills region of the northern San Fernando Valley[206] (Figure 46 in Volume I, Chapter 2). According to Scott,[206] this estimate was based, in part, on the catastrophic building damage at the Holy Cross Hospital, Olive

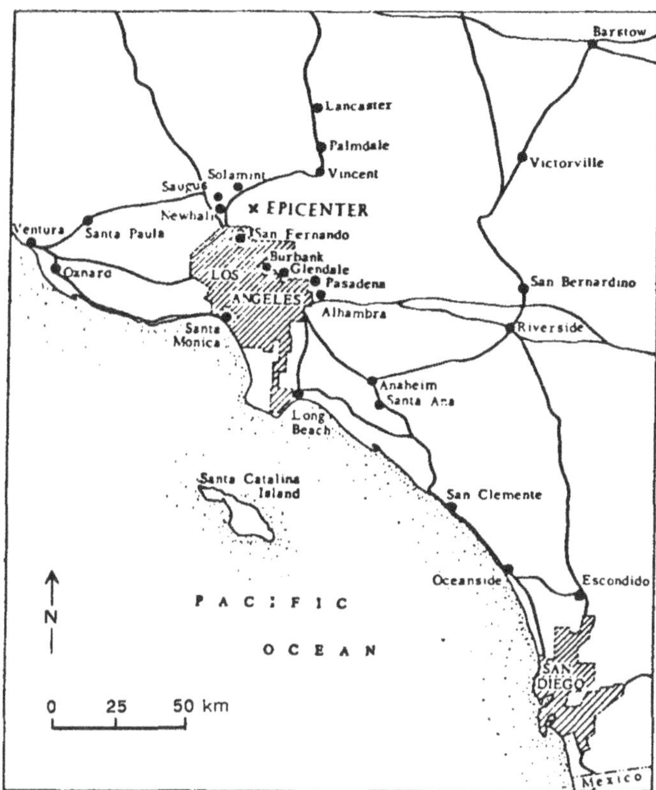

FIGURE 13. Epicenter of the February 9, 1971 San Fernando, California earthquake ($M_L = 6.4$) and its geographic relationship to southern California cities. (From Steinbrugge, K. V. and Schader, E. E., Earthquake damage and related statistics, in San Fernando, California, Earthquake of February 9, 1971, Vol. I (Part A), Benfer, N. A., Coffman, J. L., and Dees, L. T., Eds., U.S. Government Printing Office, Washington, D.C., 1973, 692.)

View Community Hospital (Figure 15 in Volume II, Chapter 3), and San Fernando Veterans Administration Hospital (Figure 16 in Volume II, Chapter 3).

The San Fernando seismic event provided the most recent test of modern earthquake-resistant construction in the U.S. However, as described by the Earthquake Research Institute Committee,[957] a number of buildings "failed the test." The response characteristics of several classes of construction and architectural elements are discussed in this section; lifeline performances are discussed in a later section of this chapter.

1. Modern Lightweight Industrial and Commercial Buildings

A modern type of lightweight industrial and commercial building to be tested by this earthquake was a single-story structure with a plywood roof and *unit masonry* (e.g., reinforced brick, reinforced hollow concrete block) or tilt up reinforced concrete walls.[994] For the latter type of wall system, the concrete panels are cast flat on the floor slab and lifted or tilted up into place, hence the name *tilt-up buildings.*[995] A tilt-up building in an early stage of construction is shown in Figure 14. Most of the industrial and commercial buildings were less than 10 years old at the time of the earthquake and were designed and constructed in accordance with the earthquake bracing provisions of the City or County of Los Angeles building codes.[995]

TABLE 2

Damage Estimate—February 9, 1971 San Fernando, California Earthquake (M_L = 6.4)[a]

Structure	Number damaged	Amount (dollars)
Schools	180	22,500,000
Hospitals	4	50,000,000
Residential	(23,570)	(179,500,000)
Homes	21,761	
Apartment houses	102	
Mobile homes	1,707	
Commercial buildings	542	
Miscellaneous structures	250	
Highways and roads		27,500,000
Dams		36,500,000
Other public structures		145,000,000
Utilities		42,000,000
Personal property		50,000,000
Total	24,546	553,000,000

[a] Estimate does not include the cost of emergency services, loss of employment, and loss of revenue from taxes and change in the tax base.

From Kachadoorian, R., The San Fernando, California, Earthquake of February 9, 1971, U.S. Geological Survey Professional Paper 733, U.S. Government Printing Office, Washington, D.C., 1971, 5.

A total of 69 damaged buildings, located between 13 and 27 km from the epicenter, were examined in detail; 63 of the buildings were located in four industrial tracts: Sylmar-3,[996-999] San Fernando-4,[1000-1004] Bradley-23, and Arroyo-33,[994] and six buildings were sited at individual sites in the northern part of the valley.[1005-1010] For the seven buildings in the Sylmar and San Fernando industrial tracts and the six isolated buildings, several similar types of damage were observed because of similarities in building materials and types of construction.[1011]

1. One of the most common types of structural failure was the detachment of the roof from the end walls (i.e., walls parallel to the roof girders). The separation was usually caused when the perimeter nails pulled through the edges of the plywood sheathing or out of the wall ledgers.* In still other cases, the ledgers failed in cross-grain bending, resulting in complete separation from a wall. The ledger-to-rafter connection was commonly a metal seat which was capable of supporting only vertical loads. Consequently, when the walls moved laterally, the adjacent portions of the roof generally collapsed (Figure 15). This type of collapse represents a serious life hazard because the failure is sudden or "brittle", offering little warning to the occupants. Similar buildings that had strap-type steel joist anchors at exterior walls, and not wood ledgers, for anchoring the walls and the roof experienced no separations during the earthquake[1012] (Figure 16). Following the earthquake, the Los Angeles City Building Code was amended to prohibit the use of wood ledgers in place of joist anchor connections.[1012]
2. The roof-to-side wall (i.e., a wall normal to and supporting roof girders) connec-

* Horizontal board attached to a wall for anchoring the roof rafters.

FIGURE 14. Precast, tilt-up concrete building under construction in Los Angeles, California. (Courtesy of James L. Ruhle and Associates, Fullerton, Calif.)

tions performed more satisfactorily because the stability of the side walls was improved by the stiffening effects of built-in pilasters (i.e., rectangular columns projecting partially from a wall). Although numerous buildings had damage in the roof-to-side wall connections, no collapses occurred.

3. Damage in the form of vertical cracks and shattering occurred at some wall corners where there was a lack of continuity of horizontal reinforcing elements.
4. Several cases of roof diaphragm separations occurred in interior areas away from the exterior walls. In these internal areas, there was a lack of continuity because the purlins (i.e., horizontal timber elements supporting wood rafters) were supported only by hangers, and the plywood sheathing was inadequate in acting as a tie across the affected buildings.

Of these 13 buildings, 11 were repaired with costs ranging from $10,000 to $136,000. Two buildings were total losses (pre-earthquake values $35,000 and $420,000), and they were demolished after the earthquake.[997-999,1001-1010] A post-earthquake view of one of the destroyed buildings is shown in Figure 17.

Of the 61 buildings in the Bradley and Arroyo industrial districts, Steinbrugge and Schader[994] examined 56 in detail. A summary of their findings is presented in Table 3. Steinbrugge and Schader estimated the earthquake damages for the 56 structures at $2,065,000, for an average damage loss of 17.7%.

2. Modern High-Rise Buildings

Most of the modern high-rise buildings in the epicentral region were medical facilities; they were all damaged.[1013] The major buildings at three hospital complexes, Olive View Community, Holy Cross, and Pacoima Memorial Lutheran, suffered severe structural damage and had to be evacuated immediately following the earthquake. The reinforced concrete buildings were constructed between 1959 and 1970 under earthquake-resistant provisions. Their sites were within a 14.5-km radius of the epicenter.[1013] The types of damage[954-962] highlighted problems such as wall movements at construction joints, pounding between adjoining wings, irregular framing systems, and complex architectural layouts.[1013] The $25 million Medical Treatment and Care Unit building at the Olive View Community Hospital was declared a total loss and was

TABLE 3

Summary of Observed Damage to Buildings Having Plywood Roofs and Unit Masonry or Tilt-Up Concrete Walls in the Bradley and Arroyo Industrial Tracts Following the February 9, 1971 San Fernando, California Earthquake (M_L = 6.4)

Element and degree of damage	Wall construction: Tilt-up	Wall construction: Unit masonry	All buildings
Concrete floor slab			
None, or hairline cracks (to 0.32 cm)	15	11	26
Moderate cracking (to 2.54 cm)	13	7	20
Severe cracking (over 2.54 cm)	7	3	10
Wall (includes five buildings for which access denied)			
None, or slight damage	11	9	20
Moderate damage to some portions	12	6	18
Severe damage to some portions	14	4	18
Collapse of some portions	2	3	5
Roofs			
None, or slight damage	11	13	24
Moderate damage to some portions	5	3	8
Collapse of some portions	20	4	24

From Steinbrugge, K. V. and Schader, E. E., San Fernando, California, Earthquake of February 9, 1971, Vol. I (Part B), Benfer, N. A., Coffman, J. L., and Dees, L. T., Eds., U.S. Government Printing Office, Washington, D.C., 1973, 706.

demolished[9, 54] (Figure 15 in Volume II, Chapter 3).* Rehabilitation costs for the Holy Cross Hospital were $4.5 million, which included the removal of the severely damaged top two floors.[962] The demolition of one wing and portions of two other wings was necessary at the Pacoima Memorial Lutheran Hospital; total damage repair costs were in excess of $1 million.[960]

At the time of the San Fernando earthquake, there were more than 190 steel and reinforced concrete-frame buildings, eight stories and above, in Los Angeles County that were constructed since 1947 under earthquake-resistant design regulations. The districts that contained high-rise construction (≥ eight stories) and their location in respect to the epicenter are shown in Figure 18. Note that the closest structures were in Panorama City, approximately 21 km from the epicenter. All of these high-rise structures survived the earthquake without serious damage to structural elements, although architectural damage was commonplace. It must be emphasized that the buildings were not fully tested because (1) of their distances from the epicenter (Figure 18) and (2) this earthquake, because of its moderate size, did not produce strong ground motion in the period range of 1 to 5 sec, a critical range of vibration for tall structures.[993,1014]

The reinforced concrete 15-story Union Bank and the 12-story Bank of California buildings in the Sherman Oaks district (Figure 18) likely had the most serious structural damage. Structural damage in the Bank of California building consisted of cracking and spalling of several concrete elements. Approximately $12,000 was spent on epoxy repair.**[1015] The Union Bank building had structural damage at the four corner columns. Costs to repair the structural damage approached $80,000.[1016]

* Structural design and earthquake damage specifics for this building are described in Section V.A of Volume II, Chapter 3.

** The epoxy technique for repairing cracked concrete is described later in this chapter.

FIGURE 15. Collapsed northwest corner of the tilt-up Warehouse Building in the Sylmar Industrial Tract following the February 9, 1971 San Fernando, California earthquake (M_L = 6.4). The cost of earthquake repairs was approximately $80,000, or 25% of the building's pre-earthquake value.[998] (Los Angeles City Department of Building and Safety photograph; courtesy of National Oceanic and Atmospheric Administration, Environmental Data Service.)

3. Public School Buildings

There were more than 9000 school buildings in the heavily shaken area, including 10 buildings within 8 km of the epicenter.[949,1017] Every building was carefully inspected after the earthquake, and the results clearly showed that the post-Field Act and renovated pre-Field Act structures performed much better than did the pre-Field Act buildings. This was especially apparent at sites where the three types of buildings stood side by side.[1017] Most of the post-Field Act and renovated buildings suffered no damage of any kind. However, nonstructural damage, usually in the form of damaged ceilings and light fixtures, cracked plaster, and displaced mechanical equipment, occurred in structures at sites where the ground motion was strong.[949] There was an absence of major structural damage in the modern buildings located in the epicentral region, which is an important life-safety finding because the ground motions in this region were probably close to the maximum values to be expected in the largest earthquakes.[1017]

Pre-Field Act school buildings at seven sites in the Los Angeles Unified School District were heavily damaged and were demolished.[949] The largest structure to be razed was the main classroom building of Los Angeles High School (Figure 19). This building was completed in 1917, 16 years before the adoption of the Field Act. It had reinforced concrete framing and unreinforced brick walls. The building was approximately 40 km south of the epicenter. The earthquake was responsible for a considerable amount of brick wall damage, and substantial portions of a masonry parapet fell through the roof of a lower wing into a stairway and classroom (Figure 20). A determination was made that it was not economically feasible to repair and strengthen the structure to meet Field Act requirements, and the classroom building was demolished at a cost of $127,000.[949,1017] The estimated cost to replace the structure was $8.2 million.[949]

4. Unreinforced Masonry Buildings

There were probably at least 40,000 pre-1933* unreinforced masonry buildings in

* Before 1933 (i.e., before the enactment of the Riley and Field acts in California), masonry buildings were constructed without reinforcing steel in their walls.[970] Consequently, this type of design considered only vertical loading and not horizontal earthquake forces.[994]

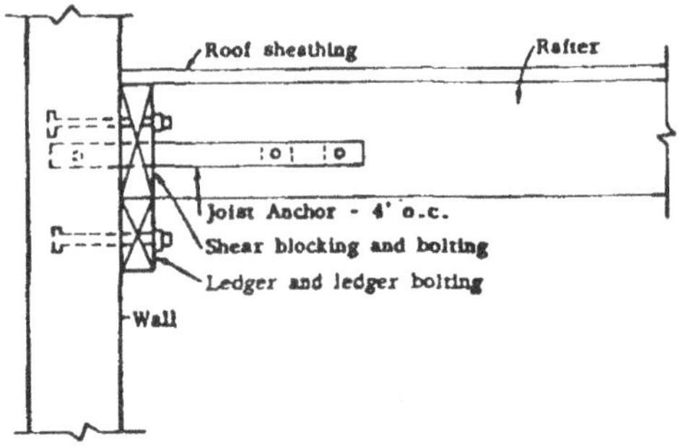

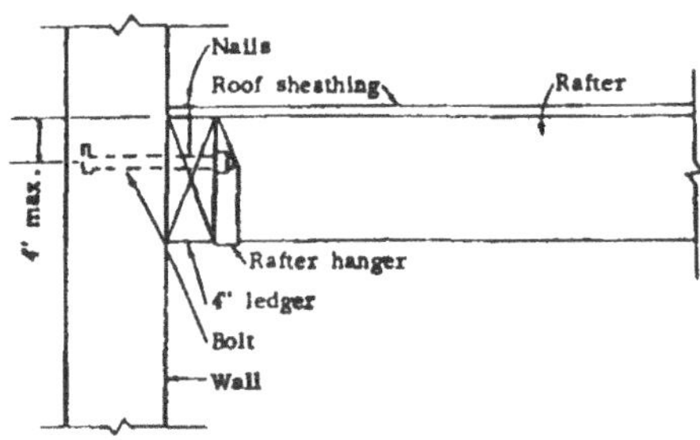

FIGURE 16. Typical joist anchorage (top) and wood ledger-roof sheathing anchorage (bottom) for roof-to-wall connections for buildings with plywood roofs and exterior unit masonry or tilt-up concrete walls. (From Briasco, E., Behavior of joist anchors vs. wood ledgers, in San Fernando, California, Earthquake of February 9, 1971, Vol. 1 (Part A), Benfer, N. A., Coffman, J. L., and Dees, L. T., Eds., U.S. Government Printing Office, Washington, D.C., 1973, 122.)

the Los Angeles metropolitan region at the time of the earthquake. Approximately 10% of these buildings were damaged or destroyed. Damage frequency varied with epicentral distance. In the City of San Fernando, located approximately 12 km south of the epicenter (Figure 5 in Volume I, Chapter 3), 75 to 85% of the unreinforced masonry buildings were severely damaged. The frequency of damage was estimated at 10% in downtown Los Angeles, approximately 40 km south of the epicenter[970] (Figure 18). The partial collapse of the Midnight Mission pre-1933 masonry building in central Los Angeles was responsible for the only death in the city due to building collapse[970] (Figure 21). At the San Fernando Veterans Administration Hospital complex, located 8 km from the epicenter in the foothills of the San Gabriel Mountains, the collapse of four pre-1933 unreinforced masonry buildings was responsible for 44 deaths (Figure 16 in Volume II, Chapter 3).

Most of the unreinforced masonry buildings were one to four stories in height, and as described by Abel,[970] construction details were simple and similar.

> Walls were made up of three wythes of brick, producing a wall thickness of 12 to 13 inches (30.5 to 32.0 cm). Outer wythes were full bedded in mortar, while the inner wythe was usually made up of culls or pieces of broken bricks . . . Mortar was sloshed into this interior space with no special effort to fill it solidly.

FIGURE 17. Collapsed southeast corner of Stone's Liquor Store in Sylmar following the February 9, 1971 San Fernando, California earthquake (M_L = 6.4). The building was constructed in 1969; it had a plywood roof and reinforced hollow concrete walls. Structural failures could have been caused by a severing of the roof-to-wall connection when the bearing walls were subjected to ground accelerations normal to their planes. The ability to resist lateral loading along the south elevation may have been further impaired because the vertical reinforcement did not appear to extend more than 1.8 m above the floor, thus leaving the top portion of the wall unreinforced. Portions of other walls with reinforcing steel did not appear to contain grout. This building, valued at $35,000 prior to damage, had to be demolished shortly after the earthquake.[997] (Courtesy of National Oceanic and Atmospheric Administration, Environmental Data Service.)

Mortar was a mixture of sand and lime but, in some instances, varying amounts of cement were added to the mix.*

Floors and roofs were framed with wood joists or nailed wood trusses, and pockets were provided in the walls for bearing of these members. At the roof, the inner wythe often was stopped at the ceiling line to form a sill. Framing would then bear on this sill, or cripple studs would be placed on the sill to raise the roof rafters above the ceiling and form an attic space . . . Girders were wood or steel beams, also set in wall pockets with walls thickened at these points to form pilasters. Often, though not always, T-bar anchors were installed at about 8 feet (2.4 m) on center to tie the framing to the wall . . . the anchor end was embedded in the wall and the other end was a 90° bend with point driven into the side of the joist and secured with a few bent nails or a wire staple. Steel beams were placed over larger openings to support the wall above . . . Other than mortar bond, there was no anchorage of the wall to the beam nor of the beam to its bearings at the opening jambs.

The sole purpose of these walls was to provide support for vertical loads, and for this purpose they served very well. The buildings were sufficiently heavy so that horizontal forces from wind could be ignored. Horizontal inertia forces from earthquakes were not a consideration.

Abel[970] describes three types of typical damage to unreinforced masonry construction by making reference to a hypothetical masonry building with the following characteristics: two stories in height, 18.4 × 36.8 m in plan, solid side walls, rear wall has

* Lime mortar deteriorates and loses strength with age and exposure. Mortar mixes were not standardized prior to 1933.[970]

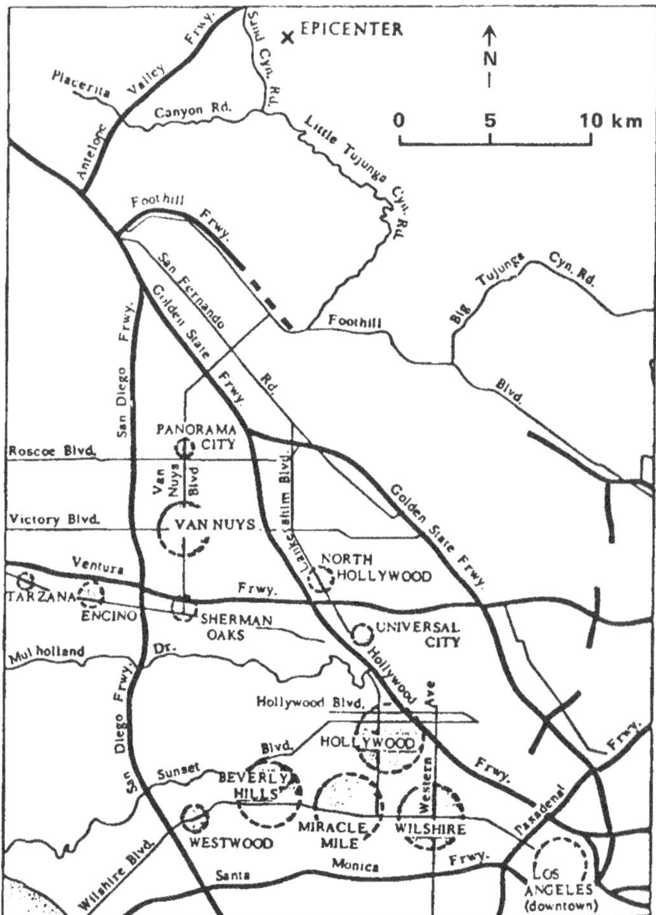

FIGURE 18. Districts containing high-rise construction (≥ eight stories) in the metropolitan Los Angeles region in 1971 and their geographic relationship to the epicenter of the February 9, 1971 San Fernando, California earthquake (M_L = 6.4). (From Steinbrugge, K. V. and Schader, E. E., Earthquake damage and related statistics, in San Fernando, California, Earthquake of February 9, 1971, Vol. I (Part A), Benfer, N. A., Coffman, J. L., and Dees, L. T., Eds., U.S. Government Printing Office, Washington, D.C., 1973, 708.)

several windows in both stories, and the front wall is mostly glass in the first story with large windows separated by narrow brick piers in the second story.

1. The rear wall moved in both transverse (i.e., perpendicular to the wall) and longitudinal (i.e., parallel to the wall) directions. Evidence for transverse motion was residual wall movement away from the building; in some cases, the movement was as much as 5.1 to 7.6 cm. In cases of extreme movement, portions of the wall collapsed. Diagonal cracks, originating from the corners of window and wall openings and indicating diagonal tension failure, were evidence for longitudinal motion.
2. Side wall damage was unusually small, although there were numerous moderate to severe parapet failures. As was described in Section V. C. of Volume II, Chapter 3, the parapet correction ordinance in Los Angeles was applicable only to cornices and appendages on walls fronting public streets or exit paths.

FIGURE 19. Los Angeles High School classroom building following the February 9, 1971 San Fernando, California earthquake (M_L = 6.4). A portion of a masonry parapet wall fell off this pre-Field Act building (between arrows). The building was demolished shortly after the earthquake. (Courtesy of National Oceanic and Atmospheric Administration, Environmental Data Service.)

3. Front walls underwent movements similar to rear walls. However, because most front walls had corrective parapet work completed prior to the earthquake, the stability of the wall was increased, thereby limiting the amount of outward wall movement. Transverse motion caused pier cracks, but they were smaller than those in the rear wall. The mostly glass first story often suffered no apparent damage. No answer is readily available to explain the lack of damage to a wall that, relative to the rear wall, had much less shear capability. It might have been that the front wall was able to move because it was flexible or that the wall was restrained by adjacent buildings.

These patterns of failure once again emphasized the need for strengthening or razing this type of hazardous construction.[992]

5. Wood-Frame Dwellings

The number of wood-frame dwellings damaged during the earthquake totaled 21,761, of which 465 (2.14%) were declared unsafe.[990] Ground-induced vibrations were responsible for damaging the vast majority of the units. Several patterns of damage in the heavily shaken area of the San Fernando Valley were as follows:

1. More recently constructed dwellings performed noticeably better than older (e.g., pre-1940) dwellings (Figure 22). Factors explaining the poorer performance of older wood-frame houses included structural deterioration due to rotting and termite damage, a lack of lateral bracing elements, and insufficient or poor anchorages between the wood sill, also called the mud sill, and the concrete or masonry foundation.[994,1018] Most dwellings built before 1940 did not have con-

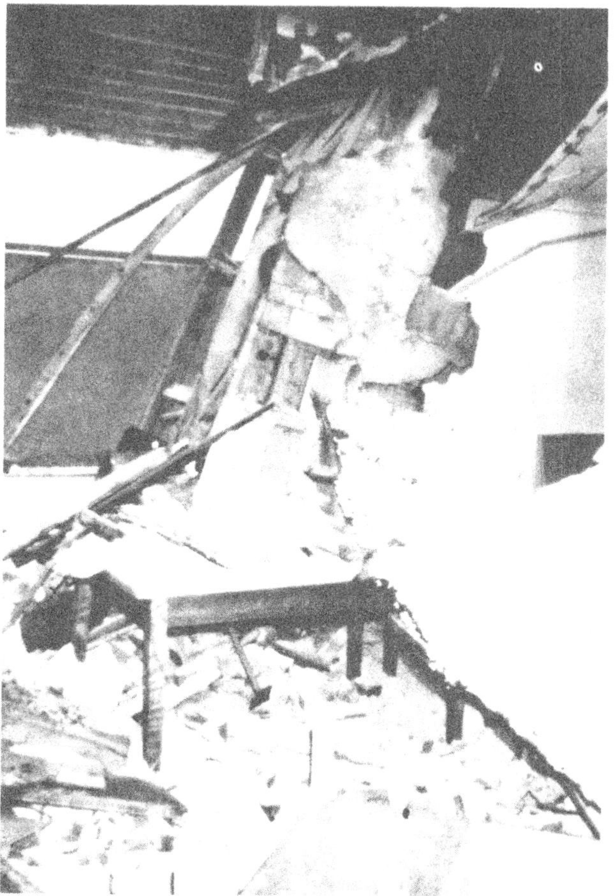

FIGURE 20. Classroom at Los Angeles High School following the February 9, 1971 San Fernando, California earthquake ($M_L = 6.4$). Portion of a parapet wall fell through the classroom roof. (D. K. Jephcott photograph; courtesy of National Oceanic and Atmospheric Administration, Environmental Data Service.)

nections between the sill and foundation, and according to Yanev,[1018] the lack of this detail is probably the single most common cause of earthquake damage to wood-frame dwellings.

2. One-story houses performed substantially better than either two-story or split-level dwellings with garages at the ground level. According to Yanev,[1018] the latter two types of buildings were especially susceptible to vibration damage because they are structurally weaker than one-story structures (subject to greater deflections caused by the weight of an upper story) and the presence of a large opening for the garage door produces, in effect, a bracing system with only three shear walls. Consequently, large earthquake-induced torsional forces were produced because of the U-shaped wall system. With the generation of torsional forces, the buildings tended to twist and because of an inadequate resistance to this type of motion, especially with the weight of a second story, walls of the garage often failed partially or totally (Figure 23). Torsional resistance was aggravated if the garage walls were inadequately braced (i.e., lack of plywood sheathing along vertical studs) for earthquake loads.[1018] According to Foth,[1019] split-level homes

FIGURE 21. Front elevation of the unreinforced masonry Midnight Mission building in downtown Los Angeles showing partial collapse of a room and second-story wall following the February 9, 1971 San Fernando, California earthquake (M_L = 6.4). Although only a 6.1-m portion of the second story wall fell, the remainder of the wall was displaced outward from 2.5 to 5.0 cm; further movement and collapse probably were prevented by the joist anchors installed as a part of the Los Angeles parapet correction program (described in Volume II, Chapter 3). One resident ran from the building at the beginning of the earthquake and was killed by falling bricks.[970] (Los Angeles City Department of Building and Safety photograph; courtesy of National Oceanic and Atmospheric Administration, Environmental Data Service.)

were also suspect to vibration damage because of "a lack of continuity of floor and roof diaphragms across the stepped portion."

3. Although not a common type of house, masonry veneered wood-frame dwellings were seriously damaged (Figure 24). This type of structure is highly susceptible to earthquake damage because the additional weight of the facing material produces larger inertia forces that can cause serious structural damage and the spalling of the veneer itself if not properly anchored to the walls.
4. Severe masonry chimney damage occurred principally in older dwellings and two-story structures[944] (Figure 25). Many chimneys built prior to 1960 did not have adequate ties to the building or reinforcing steel bars.[1018] Two-story chimneys tended to suffer more damage than one-story chimneys because of greater deflections in the second story of a dwelling with subsequent hammering against the chimney. Yanev[1018] observes that although most of the houses in the heavily shaken area were less than 15 years old and their masonry chimneys were reinforced in accordance with local building codes, approximately 32% of the chimneys suffered damage.

Because certain types of damage were common to a large number of wood-frame dwellings, amendments were incorporated into the Los Angeles City Building Code in 1972:[1020]

. . . requiring revisions in the design and construction of diaphragm sheathing, veneer ties, framing, reinforcing or concrete in masonry chimneys, anchorage of water heaters, and regulations related to cutting

FIGURE 22. Older wood-frame dwelling in San Fernando damaged during the February 9, 1971 San Fernando, California earthquake (M_L = 6.4). (Courtesy of National Oceanic and Atmospheric Administration, Environmental Data Service.)

and notching of walls and studs. It is anticipated that these changes will greatly improve the safety and stability of residential structures with a cost increase of less than 1%.

At this point, the author would like to highly recommend the book *Peace of Mind in Earthquake Country—How to Save Your Home and Life* by Peter Yanev.[1021] The book is especially useful for the individual home owner because Yanev describes (1) corrective measures in step-by-step fashion that can be made to an existing home and its contents to reduce the earthquake risk, and (2) practical measures for designing and constructing new houses in areas subject to earthquakes.

6. Mobile Homes

Mobile homes were especially vulnerable to ground motion because of foundation attachment and stability problems. Steinbrugge and Schader[994] describe the typical foundation-coach anchorage procedure that was used in southern California.

The typical setup technique is to roll the coach into place and use one of several types of piers (such as precast concrete pyramid-shaped piers) placed at intervals of about 6 feet (1.8 m) along the main frame. A screwjack on the top of each pier is then brought up into contact with the undercarriage and positioned until the coach is level and steady on the mounts. When soft soils are encountered, concrete block masonry or flat (cement) patio-type blocks are used to spread the load and prevent settling. Utilities are next hooked up, an aluminum skirt is finally applied around the base, and stairs are placed at the doorway . . . If the yolk-tongue at the forward end of the trailer is left on, . . . it is vertically supported and skirted.

Because the piers were not anchored to the ground and the screwjack levelers were not attached to the coach frame, the support technique was very unstable. During the earthquake, there were numerous cases in the meizoseismal region where coaches were shaken off their supports (Figure 26). For example, in two mobile home parks in Syl-

FIGURE 23. Partially collapsed garage section of modern split-level dwelling in Sylmar following the February 9, 1971 San Fernando, California earthquake (M_L = 6.4). (Courtesy of National Oceanic and Atmospheric Administration, Environmental Data Service.)

mar (Figure 5 in Volume I, Chapter 3) with 40 and 101 coach occupancies all units fell from their supports, and in a park in Saugus (Figure 13), located 16 km northwest of the epicenter, approximately 95% of the 92 units were dislodged from their mountings. In all, 1707 mobile homes were damaged by this earthquake (Table 2).

7. *Nonstructural or Architectural Damage*

The San Fernando earthquake serves to emphasize several important aspects of nonstructural or architectural damage that has been repeated in other recent seismic events. First, a substantial amount of the property damage, in monetary terms, can be due to nonstructural, not structural, failures; for this earthquake, the architectural damage figure was more than 50% of total damage costs.[927] This is perhaps less surprising when one considers that approximately 70% of the construction cost for a modern engineered building is for equipment and nonstructural components.[*,1022] Second, the earthquake caused considerable damage to nonstructural elements in buildings sustaining little or no structural damage.[1022] This type of damage pattern was exemplified by two, seven-story reinforced concrete Holiday Inn buildings located 21 and 42 km south of the epicenter. In the case of the building closest to the epicenter, earthquake repair costs totaled $145,000 (11% of the initial construction cost), of which $143,000 was for nonstructural repairs.[1023] For the second structure, total earthquake repair costs

* Nonstructural components are not part of the structural system and are normally added to the building during the later stages of construction.[1022] As defined by Merz and Ayers,[1022] nonstructural elements include "facades, ceilings, partitions, elevators, lights, electrical systems, plumbing, ventilation and air conditioning systems, heating systems, fire protection systems, telephone and communication systems, storage racks, and even large pieces of owner-supplied furniture or portable equipment."

FIGURE 24. Severely damaged older wood-frame dwelling with unanchored stone veneer in San Fernando following the February 9, 1971 San Fernando, California earthquake (M_L = 6.4). (Courtesy of National Oceanic and Atmospheric Administration, Environmental Data Service.)

were $95,000 (7% of the initial construction cost). Structural repair amounted to $2,500; the remainder was for repairing nonstructural damage.[1024] Third, the amount of damage sustained by nonstructural elements "could have been greatly reduced by relatively inexpensive corrective measures."[1022]

Observed nonstructural damage that represented a life safety hazard included the following examples:[979,1025]

1. Traction elevators became inoperative when generators were thrown off their mounts and control panels fell over. Guide rails also were deformed allowing counterweights (average weight 3.2 tonnes) to swing in the hoistways; this resulted in snarled cables and occasional damaged elevators cars (Figure 27). In the Los Angeles metropolitan region, 674 elevators had counterweights thrown from their guide rails, with 109 striking cars moving in the opposite direction. There were no injuries due to elevator equipment failures, largely because of the timing of the earthquake. However, several people were stranded for a short time in stalled elevators and a few were shaken when loosened counterweights struck moving cabs. All elevators in the Medical Treatment and Care Unit at the Olive View Community Hospital (Figure 15 in Volume II, Chapter 3) were rendered inoperative by the earthquake.
2. Suspended light fixtures were heavily damaged because they were free to move under the earthquake forces. Failures occurred in supporting stem hangers and chains or at ceiling connections (Figure 28). Some of the fallen fixtures weighed between 18.1 and 36.3 kg and would have been a serious life hazard had the affected buildings been occupied. Light fixture failures were very common in school buildings.
3. Glass panels failed when they were tightly mounted. By contrast, rubber-cushioned mounts prevented excess glass breakage even when the metal frames were deformed.
4. In several buildings, emergency generators attached to vibration spring mounts moved horizontally off their mounts, often severing electrical line connections.

FIGURE 25. Severely damaged chimney and roof-mounted evaporative cooler at a residence in the San Fernando Valley following the February 9, 1971 San Fernando, California earthquake (M_L = 6.4). (Los Angeles City Department of Building and Safety photograph; courtesy of National Oceanic and Atmospheric Administration, Environmental Data Service.)

The emergency generators in the Medical Treatment and Care Unit of the Olive View Community Hospital (Figure 15 in Volume II, Chapter 3) failed in this manner, rendering inoperative all electrical life-support systems.

5. Mechanical systems, such as large tanks not properly secured or with improper leg designs, either shifted or collapsed. Large steam boilers shifted up to 1.1 m, inflicting secondary damage to connected stack-pipe systems, control panels, and adjacent walls because the units were not properly attached to the floors. In many homes, hot water tanks toppled because their lightweight legs collapsed or because the tanks were not strapped to an adjacent wall.
6. There were numerous places where unanchored book racks and storage racks were toppled, and material stored on shelves was thrown to the floor.

8. Earthquake Damage Repairs

Because of the urban setting of the San Fernando earthquake, a variety of building types were damaged, and numerous materials and techniques were used for damage repairs.[1019] In the City of Los Angeles, for damaged buildings that did not conform to the code in force at the time of the earthquake, the type of repair was determined by the extent of damage. If repair costs were less than 10% of the replacement cost of the building, the structure could be repaired by original construction techniques without regard to compliance with code standards. In the 10 to 50% range of replacement cost, repair work had to conform to code specifications. The entire structure had to comply with code standards when the damage repair costs exceeded 50% of replacement costs; most unreinforced masonry buildings in this category of damage were demolished.[1019]

FIGURE 26. Mobile home in the San Fernando Valley shifted off its foundation by the February 9, 1971 San Fernando, California earthquake (M_L = 6.4). Note damage to the aluminum skirting. (Photo by Drew P. Lawrence; courtesy of National Oceanic and Atmospheric Administration, Environmental Data Service.)

This was the first earthquake where epoxy resin adhesives were used to repair cracked concrete. The technique involved the following steps:[1019]

> The pressure injection system of repairing cracks in concrete involved sealing the surfaces of the cracks with a patching compound, installing epoxy injection nipples along the cracks at a spacing of 6 to 8 inches (15.2 to 20.3 cm) on center, and then injecting the epoxy resin-hardener mixture into the cracks under pressure. The injection sequence started with the lowest level nipple into which the material was forced until it ran out of the next nipple above. Then, the lower nipple was sealed and injection proceeded with the next nipple above. This method was continued until the entire crack was injected with epoxy.
>
> The size of cracks in concrete repaired by this method varied from greater than ¼ inch (0.64 cm) down to 1/32 inch (0.08 cm) in thickness. Penetration of very fine hairline cracks was difficult. Because of the uncertainty of complete penetration into all cracks, the strength of the repaired members was assumed to be 70 percent of their original strength. In addition, cores were taken across the repaired material to verify the penetration and curing of the epoxy material. Because of the low service temperature of epoxy, repaired cracks that exceeded ¼ inch were provided with fire-resistive protection, such as plaster, over the exposed surface.

Foth,[1019] in a type of study that will hopefully become more common, describes the specific methods used in the City of Los Angeles to repair masonry chimneys and five types of buildings.

D. April 10, 1972 Fars Province, Iran Earthquake

At 5:37 a.m. local time on April 10, 1972, a devastating earthquake (M_S = 6.9) struck the agricultural region around Ghir in the Fars Province of southern Iran (Figure 29). The earthquake and its aftershocks occurred within the 300 km wide seismic zone that coincides with the folded belt of the Zagros Mountains and the Zagros thrust

FIGURE 27. Derailed counterweight struck top of elevator car in the Union Oil building in central Los Angeles during the February 9, 1971 San Fernando, California earthquake (M_L = 6.4). (Los Angeles City Department of Building and Safety photograph; courtesy of National Oceanic and Atmospheric Administration, Environmental Data Service.)

fault zone that borders it to the north.[1026] These features and the related seismic activity are thought to be a result of compression resulting from the northward drift of the Arabian plate against the Eurasian plate (Figure 48 in Volume I, Chapter 2).[413,417,1026] The earthquake occurred on an unnamed reverse fault; Dewey and Grantz[1026] could find no evidence of surface fault displacement.

Dewey and Grantz[1026] report that the earthquake killed more than 5000 of the 28,800 inhabitants (>17.4%) of the meizoseismal region who lived in permanent dwellings and injured approximately 1400.* The high percentage of death was attributable to the collapse of traditional mud-brick and mud-and-stone, heavy-roofed dwellings that are found throughout Iran.[1027,1028] Casualties were primarily women and children because many of the men were already working in the nearby fields when the earthquake struck. More than 30,000 people were left homeless by the earthquake.[1029]

The affected population was concentrated in two northwest-southeast trending val-

* Many of the inhabitants were tent-dwelling nomadic tribesmen who generally escaped the earthquake without injury.[1026]

FIGURE 28. Pendant-mounted light fixtures that were torn loose from their ceiling support base plates at the Herrick Avenue Elementary School during the February 9, 1971 San Fernando, California earthquake (M_L = 6.4). (Los Angeles City Department of Building and Safety photograph; courtesy of National Oceanic and Atmospheric Administration, Environmental Data Service.)

leys, each measuring approximately 7 by 20 km, and separated by a 400-m high ridge. Destruction of buildings was essentially complete throughout this region that encompassed Ghir, the largest village, and the village groups in the Karzin and Afzar areas (Figure 29).[1027] Hardest hit was Ghir; 3069 of its 4068 citizens (75.4%) were killed.[1026]

Most of the traditional houses were one-story high with adobe brick or stone set in mud-mortar bearing walls. The flat roofs were composed of wood poles covered with sticks and straw and veneered with 30 cm or more of mud. The wood beams were only 10-to 12-cm thick and were spaced every 30 to 40 cm atop the bearing walls; the beams rarely passed completely through the thick walls. Total roof thickness varied from 30 to 60 cm, resulting in a tremendous weight.[1026,1027] According to Dewey and Grantz,[1026] the combination of weak bearing walls and heavy roofs in these structurally stiff dwellings made them especially vulnerable to damage from strong, high-frequency accelerations. Such ground motions were responsible for wall failures or the unseating of the support beams and consequent roof collapse in virtually every traditional structure in the most severely shaken area (Figure 30).[1027]

Destruction to traditional dwellings sited on alluvium decreased markedly beyond about 15 km from the earthquake source region and was negligible beyond 27 to 40 km. According to Dewey and Grantz,[1026] such a highly localized damage pattern is often attributable to a shallow-focus earthquake with body waves, and not surface waves, being the cause of destruction. At Jahrom, located approximately 50 km east of Ghir (Figure 29), the lower frequency surface waves caused alarm, but little damage to traditional construction.[1028]

There were a few modern buildings (i.e., constructed with modern materials) in the meizoseismal region. They were usually one-story brick structures with sand-lime ce-

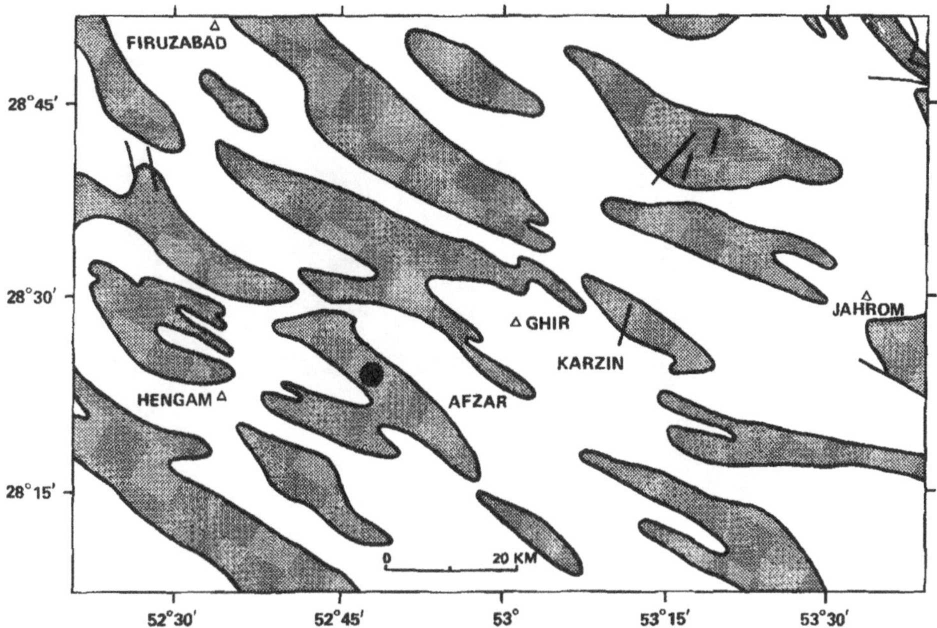

FIGURE 29. Ghir-Karzin-Afzar region of the destructive April 10, 1972 Fars Province, Iran earthquake (M_s = 6.9). Solid circle represents the epicenter; mountain ranges are shaded and known pre-earthquake faults are shown as solid lines. (From McEvilly, T. V. and Razanil,R., *Bull. Seismol. Soc. Am.*, 63, 341, 1973. With permission.)

ment mortar and roof systems comprised of steel I-beams, spaced 80 to 100 cm apart, with interbeam slabs formed by shallow brick arches.[1027] These buildings fared little better than the traditional structures (Figure 31). The primary types of destruction were the collapse of bearing walls that failed because little or no reinforcing steel was used and roof collapse caused by the separation of the roof beams from the bearing walls. Separation occurred because of inadequate ties to the walls.[1027,1028] McEvilly and Razani[1027] note that the structural inadequacies were created largely because the local, unskilled workers were unfamiliar with modern materials and construction techniques.

E. September 6, 1975 Lice, Turkey Earthquake

At 12:22 p.m. local time on September 6, 1975, a destructive earthquake struck near the town of Lice in the eastern Anatolia region of Turkey (Figure 32). The U.S. Geological Survey (USGS) established its surface wave magnitude (M_s) at 6.7.[1030] The earthquake occurred on an unnamed thrust fault at a depth reported as 16 km by Turkish authorities and as 33 km or shallower by the USGS. The earthquake was felt over an area of 210,000 km². Although only of moderate size, the earthquake killed at least 2386 people and injured an additional 4500 citizens. More than 5275 dwellings were destroyed and another 6850 units were damaged; property losses were estimated at $17 million (U.S.).[1031,1032]

The town of Lice and its surrounding villages, located approximately 5 km from the epicenter, were within the most severely affected area (Figure 32). In the Lice area there were 1312 fatalities out of a population of about 8200 (16%). Of 6760 dwellings, 6284 suffered some type of damage (93%); of these, 4982 collapsed or sustained heavy damage—a 73.7% loss. In Hani and its villages, located approximately 27 km from the epicenter (Figure 32), there were 112 deaths and 295 out of 3135 dwellings collapsed or sustained heavy damage.[1031] As reported by Yanev,[1031] both Lice and Hani are di-

A

B

FIGURE 30. Collapsed traditional dwellings in the villages of Tang Rudeh (A) and Shar-e Pir (B) following the April 10, 1972 Fars Province, Iran earthquake (M_S = 6.9). (Courtesy of James W. Dewey, U.S. Geological Survey.)

FIGURE 31. The steel beam masonry Malaria Eradication Center building in Ghir following the April 10, 1972 Fars Province, Iran earthquake (M_s = 6.9). Note the absence of reinforcing steel in the bearing wall. (Courtesy of James W. Dewey, U.S. Geological Survey.)

vided into old and new sections. The old sections are situated on the steep foothill slopes of the Taurus Mountains and contain most of the traditional stone houses. The newer-town sections and most of the larger structures occupy the lower, more level areas.

P. I. Yanev surveyed most of the damaged areas 3½ weeks after the earthquake. The following discussion is an overview of Yanev's damage assessments for several types of construction:[1031,1032]

1. Yanev was able to examine two damaged reinforced concrete buildings in Lice (Figures 33A and B and 34). Both structures had reinforced concrete frames and stiff, unreinforced brick infill walls. The brick walls represent nonstructural elements, and their principal effect is to increase the stiffness of the surrounding concrete frame (i.e., frame is less able to absorb earthquake forces by bending). With the presence of infill walls, seismic loads become concentrated at the infill bays, and within a bay, the total shear force must be resisted by each column and not divided between the two columns as would be the case if the lateral load was resisted by frame bending.[1033] Frame failures in the two buildings were most common at the tops and/or bottoms of the columns where an infill wall had failed in shear. This type of reinforced concrete building has performed similarly in several other contemporary earthquakes,[1033-1037] and emphasizes how nonstructural elements can drastically alter the predicted behavior of the structural frame.[1033]
2. There were several dozen two- and three-story unreinforced masonry buildings in Lice and Hani at the time of the earthquake. The buildings had external bearing walls of unreinforced brick or cut stone held together by weak mortar with a high lime content and reinforced concrete floor slabs. Most of the buildings in

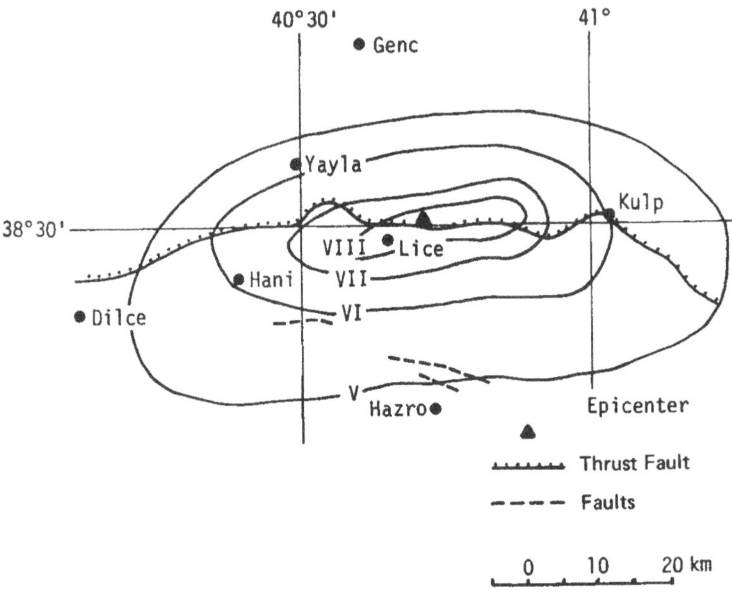

FIGURE 32. Isoseismal map of the September 6, 1975 Lice, Turkey earthquake (M_s = 6.7). See Table 8 in Volume I, Chapter 2 for Modified Mercalli Intensity definitions. (From Yanev, P. I., *The Lice, Turkey, Earthquake of September 6, 1975 (Reconnaissance Report)*, URS/John A. Blume & Associates, San Francisco, Calif., 1975. With permission.)

Lice suffered severe damage, and several experienced partial collapse. The police station and two high school buildings suffered spectacular vibration-induced damage and were declared total losses (Figures 35 and 36). The external bearing walls of these three structures sustained severe shear failures. Yanev speculates that the buildings did not collapse because they were being held up by interior unreinforced brick walls (interior inspections were not possible for safety reasons). In several cases, portions of corner walls were destroyed by torsion and pounding between the floor slabs and walls (Figure 35). Unreinforced masonry chimneys were commonly broken off at roof lines, often falling through the wood and tile roofs (Figure 35).

The hospital building in Lice was the only unreinforced brick building to escape serious damage (Figure 37). It was located in the older, foothills section of town, whereas the police station and high school buildings were sited on alluvium at the base of the foothills. Yanev attributes the limited damage to the hospital building to lower intensities of ground motion rather than to superior construction.

3. Most of the buildings damaged or destroyed by the earthquake were stone dwellings of a style common to eastern Turkey. The walls are typically comprised of broken pieces of stone held together with mud, or less commonly with cement mortar. Small stones are often keyed in between larger stones. The flat roofs are composed of about 20 support timbers and 40 to 60 cm of mud. This type of roof is very popular because it provides good insulation against the hot summers and cold winters of Anatolia, is watertight with regular maintenance, and serves as a platform for storing firewood and hay. However, the roofs may weigh as much as 1000 kgm/m^2, and their collapse appears to have been responsible for most of the casualties (Figure 38). Roof failure was usually caused by the with-

A

B

FIGURE 33. (A) The new reinforced concrete municipal building in Lice following the September 6, 1975 Lice, Turkey earthquake (M_s = 6.7). The building suffered moderate damage to the frame and significant damage to brick infill walls. (B) Interior of the entrance hall of the municipal building showing damage to the reinforced concrete frame and brick infill walls. (Courtesy of Peter I. Yanev; URS/John A. Blume & Associates, San Francisco.)

drawal of support timbers from the walls. These stone houses, which make up 92% of all dwellings in the Lice-Hani region, were usually total losses in areas having Modified Mercalli Intensities of VIII or higher (Figure 39).

4. Following the destructive May 22, 1971 Bingöl earthquake (M_L = 7.0) that struck 50 km north of Lice, the Ministry of Reconstruction and Resettlement (MRR) of the government of Turkey provided earthquake-resistant single family dwell-

FIGURE 34. A reinforced concrete building in Lice that suffered several column shear failures (lower right corner) and numerous infill wall failures as a result of the September 6, 1975 Lice, Turkey earthquake (M_s = 6.7). (Courtesy of Peter I. Yanev; URS/John A. Blume & Associates, San Francisco.)

ings for the refugees that were designed by the MRR's Earthquake Research Institute (ERI). A number of these structures were built in Lice and Hani for resettlement purposes. The houses have wood frames and prefabricated wall panels with interior insulation. A wall panel has a wooden frame covered by a cement board on its exterior side.* Several of these earthquake-resistant dwellings were located in an otherwise heavily damaged area of Lice, but none was seriously damaged; most commonly clay tiles were shaken off the roofs. In Hani, the houses did not have tile roofs, and no damage was noticeable (Figure 40). Because this type of dwelling survived the earthquake without causing casualties, the MRR constructed approximately 3000 similar units throughout the stricken area (Figure 41).

IV. LIFELINE RESPONSES TO EARTHQUAKE HAZARDS

Urban areas have a number of lifeline systems that Duke and Moran[1038] categorize as follows:

Energy	Water
Electricity	Potable
Gas	Flood
Liquid fuel	Sewage and solid waste

Transportation	Communication
Highway	Telephone and telegraph
Railway	Radio and television
Airport	Mail and press
Harbor	

* The dwellings cost about $1300 in 1971; they could be purchased over a 20-year period with no interest charge.[1031]

FIGURE 35. Unreinforced masonry police building in Lice following the September 6, 1975 Lice, Turkey earthquake (M_s = 6.7). Note the severe damage to the first floor corner of the building, the shear failure of the piers between the windows, and the toppled brick chimneys. (Courtesy of Peter I. Yanev; URS/John A. Blume & Associates, San Francisco.)

Most lifelines consist of "sources, major transmission lines, storage, and a distribution or collection system...Each has a terminus outside the city and an extensive matrix of contact or distribution points inside."[1038] Lifelines represent approximately 50% of the economic value vulnerable to earthquakes in an urban area.[1039] As noted by Duke and Moran,[1039] earthquake experience has demonstrated repeatedly that lifeline failures can produce severe consequences to public health and welfare; consequential examples include: "(1) loss of service of the utility; (2) direct financial loss; (3) suspension of human activity, e.g., employment; (4) an inability to cope with secondary disasters such as fires, famines, and epidemics; and (5) failure of a nature such that a lifeline itself becomes a hazard to life and property."[1038]

More specifically, essential lifeline interruptions can have immediate and serious effects upon a population.

1. Damaged transportation systems can impede evacuations or the arrival of disaster relief personnel and supplies.
2. Ruptured gas lines and severed electrical cables can be catalysts for fires.
3. Damaged water lines, storage tanks, and aqueducts can hamper fire fighting efforts and make potable water a rare commodity.
4. Ruptured sewer lines, municipal sewage tanks, and septic tanks can contaminate drinking water and render home toilets inoperative.
5. The interruption of gas and electrical service can make it difficult to heat or cool buildings, prepare food, or boil water.
6. Downed telephone lines or damaged equipment can make it difficult for people of a stricken area to contact relatives and vice versa or to immediately contact emergency relief agencies.

Past experience has shown that it can take weeks or even months to fully restore damaged lifeline services and years for the complete restoration of a destroyed source facility or distribution network. The repair/replacement time is not only a function of absolute damage, but also of the economic and technological level of the society in question.

FIGURE 36. West-wall view of an unreinforced masonry high school building in Lice following the September 6, 1975 Lice, Turkey earthquake (M_s = 6.7). All lateral resisting elements were shattered. Similar damage occurred to an adjacent high school building of identical construction. (Courtesy of Peter I. Yanev; URS/John A. Blume & Associates, San Francisco.)

A. Lifeline Performance—February 9, 1971 San Fernando, California Earthquake

Throughout the world, earthquake-resistant engineering concepts have been directed primarily toward buildings, especially high-rise structures.[1040] In the U.S., seismic building criteria were first established following the March 10, 1933 Long Beach, California earthquake (M_L = 6.3). By contrast, the February 9, 1971 San Fernando, California earthquake (M_L = 6.4) had the greatest impact on the development of seismic design criteria for vital lifeline systems. This was the first time that a major effort was directed towards investigating and preparing detailed reports of lifeline performances during an earthquake. The monumental task was completed by members of three special subcommittees, Energy and Communication Systems,[1041] Water and Sewage Systems,[1042] and Transportation Systems,[1043] of the National Oceanic and Atmospheric Administration/Earthquake Engineering Research Institute (NOAA/EERI) Earthquake Investigation Committee.[1044] The following discussion is concerned with the response of several lifeline systems to the strong ground motion and differential earth displacement hazards associated with the San Fernando earthquake.

1. Energy and Communication Systems

The earthquake damaged a number of electrical facilities operated by the Los Angeles Department of Water and Power (LADWP) and the Southern California Edison Company (SCE), which disrupted power service to large portions of the Los Angeles Basin.[1045,1046] The most severe damage to LADWP facilities occurred within a 4.8-km radius of Sylmar (Figure 5 in Volume I, Chapter 3), an area where peak ground accelerations were estimated to have been between 0.3× and 0.5× g. Within this zone, the heaviest damage occurred at the Sylmar Converter Station (40% equipment loss), the Olive Switching Station (80% equipment loss), and the Sylmar Switching Station (90%

FIGURE 37. South-wall view of the unreinforced masonry hospital building in Lice following the September 6, 1975 Lice, Turkey earthquake (M_s = 6.7). This was the only unreinforced masonry structure in Lice to have escaped serious structural damage. (Courtesy of Peter I. Yanev; URS/John A. Blume & Associates, San Francisco.)

equipment loss) (Figure 42). The financial loss at the Sylmar Converter Station alone was $22 million, with a restoration time estimated at 1.5 to 2 years.* Ground shaking caused most of the damage at this station, but permanent ground movements contributed to the damage in certain cases.[1045]

Outdoor (yard) equipment at the above facilities was especially vulnerable to the earthquake forces. Damage occurred when certain pieces of equipment, such as circuit breakers, transformers, and air switches, were toppled from their foundation pads or pedestals as a result of anchoring system failures (Figure 42). Other pieces of equipment, including conductors and condensers, were heavily damaged when supporting porcelain insulators failed during the earthquake.[1045,1047]

* Youd[495] notes that this facility is particularly important "because it ties southern California to the Bonneville electrical power system, and is the facility that converts the power from direct current to alternating current."

FIGURE 38. Interior view of a traditional stone dwelling in Lice showing collapsed timber and mud roof following the September 6, 1975 Lice, Turkey earthquake (M_S = 6.7). (Courtesy of Peter I. Yanev; URS/John A. Blume & Associates, San Francisco.)

erations varying from 0.1× to 0.25× g; only minimal damage was reported for certain pieces of equipment.[1046]

Other types of LADWP facilities and equipment were damaged in the area of the most intense ground motion. For example, the San Fernando Powerplant suffered severe structural damage from intense shaking and ground settlement. The reinforced concrete powerhouse was constructed in 1921 without earthquake-resisting elements. In addition, approximately 285 overhead distribution transformers and 30 wooden poles were damaged and had to be replaced. Many lattice-type steel towers were subjected to intense shaking and permanent ground movements, but all transmission lines remained intact even though about 20 towers were displaced from their original positions.[1045] As noted by Wong,[1045] the towers had an inherent earthquake-resistance because the wind and other loads used in their design exceeded the seismic loading.

Power failures, resulting from damaged equipment and relays triggered by the earthquake, affected 636,000 LADWP and 254,000 SCE customers.[1045,1047] Power was restored to all customers by February 12—a maximum interruption of 3 days.[1046] Damage patterns revealed that full-scale testing of some equipment under simulated earthquake loading is needed.[1047]

More than 500 communities (3.1 million customers) are supplied with natural gas by the Southern California Gas Company; an affiliate, Pacific Lighting Service, operates transmission lines throughout most of southern California. Four transmission lines that deliver gas from the San Joaquin Valley to the Los Angeles Basin were damaged between Newhall and San Fernando (Figure 13). These welded steel lines were shut in immediately after the earthquake which resulted in a supply loss to the distribution system in the San Fernando/Sylmar area (Figure 5 in Volume I, Chapter 3). A total of 68 breaks had to be repaired to fully restore service in the four lines; the damage was repaired by February 12.[1048]

A 28- to 31-km² area in the northern San Fernando Valley, which included the communities of Sylmar and San Fernando, was the only region where the gas distribution system suffered serious damage.[1048] There, violent earth movements extended, com-

FIGURE 39. View of a traditional stone dwelling in Lice following the September 6, 1975 Lice, Turkey earthquake (M_s = 6.7). Total roof collapse is evident at photo (right); note pile of roof support timbers. (Courtesy of Peter I. Yanev; URS/John A. Blume & Associates, San Francisco.)

pressed, and twisted the piping system that in turn caused broken mains, valves, and service risers. Approximately 450 breaks were discovered. It is thought that no building fires were associated with the breaks. Gas service to approximately 17,000 customers was restored in most part by February 20—an 11-day interruption.[1047,1048]

The most serious damage to a General Telephone Company of California (GTC) facility occurred at the Sylmar Central Office building. There, 91 tonnes of automatic switching equipment, mounted on vertical bays, was destroyed—a loss of $4.5 million. The switching equipment was damaged or destroyed when the supporting structures failed and toppled to the floor. The design criterion for the equipment superstructure was 0.2 × g for lateral forces. The loss of this switching station severed telephone service for 9500 customers in Sylmar. Although it was not until March 19, 39 days after the earthquake, that normal service was fully restored, emergency public telephones and an emergency message center were set up next to the Sylmar Central Office building the day after the earthquake.[1048] In refurbishing the Sylmar facility, GTC strengthened the mechanical connections to insure that the equipment-supporting structures had the same resistance to lateral loading as the bulding.[1050]

2. *Water Supply and Sewerage Systems*

Water supply and sewerage systems also were hardest hit in the northern San Fernando Valley. There, two hydraulic fill dams were severely damaged and incidents of damage to aqueducts, tunnels, storage tanks, pipelines, water wells, and house connections for water and sewage were reported.[495,1050-1052] The damaged Upper and Lower San Fernando dams experienced partial collapse and dramatized the seismic hazard potential when a dam is sited within a metropolitan region (Figure 43). The dams and their impoundments, Upper and Lower Van Norman reservoirs (Figure 5 in Volume I, Chapter 3) are part of the Los Angeles Aqueduct System which supplies part of the water needs of the Los Angeles metropolitan region. The lower dam was constructed between 1912 and 1915 by full-hydraulic fill methods, and in later years it was enlarged

FIGURE 40. Undamaged earthquake-resistant dwellings in Hani following the September 6, 1975 Lice, Turkey earthquake (M_s = 6.7). (Courtesy of Peter I. Yanev; URS/John A. Blume & Associates, San Francisco.)

by semihydraulic and fill compaction methods. The upper dam was constructed between 1919 and 1921 by semihydraulic methods.[1052]

Upper San Fernando Dam, immediately above Lower Van Norman Reservoir (Figure 5 in Volume I, Chapter 3), subsided 0.9 m and shifted 1.5 m downstream as a result of the earthquake. Portions of the concrete lining on the upstream slope were broken and displaced, and the downstream embankment was cracked in numerous places.[495,1052] According to Youd and Olsen,[495] possible causes of subsidence and the downstream movement of the dam include "seismic compaction, lateral spreading of the embankment, and foundation movements associated with landslides" on slopes near the dam.

The principal damage to the Lower San Fernando Dam was a spectacular upstream slope failure that dislodged a major portion of the embankment and concrete dam face, depositing them on the reservoir floor (Figure 5 in Volume I, Chapter 1). The length of the slope failure was approximately 550 m. The slope failure reduced the dam height by 9 m. Had not the reservoir level been 10.7 m below the crest at the time of the earthquake, the dam would have been overtopped, with the likely loss of the dam and a residential area of 80,000 people would have been inundated[1053] (Figure 43). According to Youd and Olsen,[495] the slope failure could have been caused by "seismically induced inertia forces alone or in concert with liquefaction of the embankment materials, tectonic deformation, or foundation soil failures." Some 80,000 residents below the dam were evacuated for 4 days while the reservoir was lowered to a level that would preclude flooding if the dam was further damaged by aftershocks.[495] The dam was permanently taken out of service.[1052]

In the vicinity of the Upper and Lower San Fernando dams, sections of several concrete aqueducts were damaged to such an extent that they were unusable until major repairs were made. Broken and cracked concrete linings and displaced aqueduct banks were common types of damage and were usually caused by slumping of an aqueduct or displacement of the structure by an underlying slope failure.[495] Damage to two of the aqueducts disrupted service to approximately 10,000 customers in Los Angeles for a short time.[1050]

The City of San Fernando's water supply and distribution system was devastated by the earthquake, and service to 17,000 residents was interrupted.[1051] The city's supply of water was from seven wells north of the city. Four of the wells sustained repairable damage, and one was damaged beyond repair. The major impact of the earthquake on the wells was their contamination resulting from many broken sewer lines and septic tanks in the immediate area.[1052] The distribution system was so severely damaged by

FIGURE 41. Earthquake-resistant house under construction in Lice 3½ weeks after the September 6, 1975 Lice, Turkey earthquake (M_s = 6.7). The house has a wood frame and prefabricated wall panels with interior insulation. A wall panel has a wooden frame covered by a cement board on its exterior side. (Courtesy of Peter I. Yanev; URS/John A. Blume & Associates, San Francisco.)

tectonic and nontectonic ground displacements and violent ground shaking that a new system was required for the city.[1053] Immediately following the earthquake, tanker trucks supplied water for basic needs and a temporary water system was later installed above grade.[1054] The temporary distribution system used steel pipe placed along street gutters; from these lines, taps were used for hose connections to service the homes.[1052]

The NOAA/EERI Subcommittee on Water and Sewerage made a number of recommendations to reduce earthquake damage to water and sewerage systems; several of the recommendations include:[1052]

1. Modern earthfill dams that were well compacted and that rested on a moderately dense alluvial foundation were not seriously damaged by severe shaking or tectonic disturbances. By contrast, old hydraulic fill dams were severely damaged. The dynamic stability of all hydraulic dams should be investigated. This is of paramount importance for dams sited upstream from populated areas.
2. All potential dam sites should be explored by test trenches to determine if fault traces exist in the alluvium from previous earthquakes. Sites where faults exist must be avoided for dam sites.
3. Steep slopes should be avoided for the placement of pipelines due to the potential of seismically induced landslides.
4. Some type of slip joint should be used on water well pump mountings and casings to minimize twisted pump stems and damaged casings due to lateral and vertical displacements.
5. Regarding the postearthquake contamination of water wells, all septic tanks and cesspools within a specified distance from a water well should be abandoned, cleaned, and backfilled when other methods of sewage disposal are available. A well should be abandoned if no other source of disposal is available.
6. Additional reinforcing steel is needed in concrete aqueducts and tunnel linings to avoid fracturing where faults are crossed.
7. No known material or design could have resisted the magnitude of the earthquake forces that damaged water mains and fittings in the meizoseismal region. However, the use of relatively flexible joints, fittings, and pipe material offers the

FIGURE 42. Portion of the damaged 230-kV switchyard at the Sylmar Switching Station following the February 9, 1971 San Fernando, California earthquake (M_L = 6.4). Note toppled transformer at A; 7 of 12 transformers were toppled from their steel support pedestals. Damaged circuit breakers are shown at B; major damage was sustained by all 11 of the station's circuit breakers.[1045] (Courtesy of James L. Ruhle & Associates, Fullerton, Calif.)

most practical solution for a distribution system to be earthquake-resistant in areas removed from the largest ground displacement. Consequently, lead-caulked and rubber gasket joints on cast iron water mains provide greater flexibility than cement-caulked joints.

8. Necessary crossings of faults by supply systems should be above ground so that potential damage can be repaired quickly. This practice was used by the California Department of Water Resources when the California Aqueduct crossed the North Garlock, Garlock, and San Andreas faults.
9. There is no substitute for well-equipped and well-trained repair crews to quickly restore service in damaged pipelines and related facilities.

Because the earthquake nearly caused the total failure of Lower San Fernando Dam, a dynamic type of analysis was developed by scientists at the University of California, Berkeley for portraying the potential performance of earthen dams during earthquakes. By this procedure, the dynamic response of a dam is carried out by using accelerograms of real and simulated earthquakes as input for time-dependent analyses; it more accurately depicts the performance of earthen dams than the earlier pseudostatic method of analysis. Under the jurisdiction of the Division of Safety of Dams (DSD) of the California Department of Water Resources, all earth dams under state jurisdiction must be analyzed by the dynamic analysis technique. Dams found to be incapable of resisting certain earthquake forces must be rehabilitated or abandoned.

FIGURE 43. Damaged Lower San Fernando Dam one day after the February 9, 1971 San Fernando, California earthquake (M_L = 6.4) and its relation to the densely built-up area downstream. The complex at the left edge of the photograph is a school. (National Aeronautics and Space Administration photograph; from Rush, M., Holguin, A., and Vernon, S., *Potential Role of Remote Sensing in Disaster Relief Management,* School of Public Health, University of Texas, Houston, no date, 49A. With permission.)

Dynamic analyses are not yet required for concrete dams because of greater complexities involved in interpreting the analytic results.[1055]

Damage to dams was also responsible for legislation passed in 1973 that requires inundation maps to be prepared by dam owners* and approved by the State Office of Emergency Services (OES). The maps delineate areas that are likely to be flooded in

* Although the legislation applied only to dams under state jurisdiction, federal agencies volunteered to prepare inundation maps for their projects.[1055]

the event of dam failure. Once OES approves an inundation map, local communities are then required to prepare evacuation plans.[1055]

3. Transportation Systems

A total of 62 bridges were damaged by the earthquake. Of these, 25% collapsed or were severely damaged, 50% were moderately damaged, and 25% suffered relatively minor damage. Highway bridges were hardest hit within a belt 8 km long and 10 to 16 km southwest of the epicenter along Interstate Routes 5, 210, and 405 in the San Fernando area. Structures along a portion of State Route 14 near Solemint, 8 to 11 km northwest of the epicenter, were moderately damaged. Of the 62 damaged bridges, 58 (93.5%) were located within these two areas.[1056] Interstate Route 5 was blocked to traffic in both directions by collapsed bridges. Seismic shaking appears to have been the primary cause of bridge damage.[495]

The heaviest concentration of damage occurred at two freeway-to-freeway interchanges—Interchange 5/210 and Interchange 5/14 (Figure 44).[1056,1057] The interchanges are approximately 1.6 km apart along Interstate 5 and 12 km southwest of the epicenter. All 15 bridges at the Route 5/210 Interchange sustained damage ranging from cracked and spalled concrete to complete failure. From this total, two bridges experienced partial collapse and two total collapse (Figure 44). Damage was less extensive at the Route 5/14 Interchange. Of 17 structures, many of which were under construction at the time of the earthquake, one experienced partial collapse and nine sustained light to moderate damage.[1056] Bridge footings at the 5/210 Interchange were sited in alluvium, whereas footings at the 5/14 Interchange were anchored in well consolidated sandstone. The difference in foundation materials could have accounted for greater movements of bridge structures at the 5/210 Interchange, resulting in more extensive damage.[1056,1058]

Freeway bridges were designed and constructed to resist lateral earthquake forces in accordance with the following equations:[1056]

$$EQ = KCD \qquad (1)$$

where EQ = earthquake force applied horizontally at the center of gravity of the structure; the force shall be distributed to supports according to their relative stiffness. K = numerical coefficient representing energy absorption of the structure:

- K = 1.33 for bridges where a wall with a height-to-length ratio of 2.5 or less resists horizontal forces applied along the wall.
- K = 1.00 for bridges where single columns or piers with a height-to-length ratio greater than 2.5 resist the horizontal forces.
- K = 0.67 for bridges where continuous frames resist horizontal forces applied along the frame.

C = numerical coefficient representing structure stiffness, determined by:

$$C = \frac{0.05}{\sqrt[3]{T}} \qquad (2)$$

(maximum value of C = 0.1)

T = period of vibration of the structure, determined by:

$$T = 0.32\sqrt{D/P}$$

FIGURE 44. Portion of the Route 5/210 Interchange following the February 9, 1971 San Fernando, California earthquake (M_L = 6.4). The interchange had 15 bridge structures and all sustained damage ranging from cracked and spalled concrete to total collapse. Collapsed bridges blocked Interstate Route 5 in both directions and the main lines of the Southern Pacific Railroad (foreground). Two men were killed when a bridge span (arrow) fell on their truck.[495,1056] (Courtesy of James L. Ruhle & Associates, Fullerton, Calif.)

for single-story structures, where D = dead-load reaction of the structure and P = force required for a 1-in. (2.54 cm) horizontal deflection of the structure. Calculated EQ could never be less than 0.02D, and special consideration was given to bridges constructed on soft materials and to bridges having massive piers. Earthquake coefficients (EQ) for freeway bridges ranged from 0.2D to 0.13D, depending upon structural conditions. Following the earthquake, interim earthquake design criteria were introduced that doubled EQ for bridges on spread footings and increased it by 2.5 times for bridges on pile footings.[1056]

According to Elliott and Nagai,[1056] the types of damage revealed several inadequacies in detail that reduced the earthquake-resistance of many of the bridges. For example, the pattern of damage to concrete columns indicated that additional reinforcing steel is needed to increase their shear capacity and to provide confinement to insure the integrity of the concrete core and an ample ductility to prevent collapse when the columns are subjected to stresses beyond yield. Another detail that reduced earthquake resistance was that on long ramps, span hinges opened up considerably during the earthquake because there were no horizontal ties to hold the sections together. In one case, a hinge opened to such an extent that it caused the collapse of a ramp structure. Longitudinal restrainers would serve to make ramps act more as a cohesive unit, thereby increasing their resistance to earthquake forces. Other inadequacies and possible remedies for future bridge construction are described by Elliott and Nagai[1056] and Meehan.[1058]

Pavement settlement at bridge approaches was common throughout the northern San Fernando Valley.[495,1057] The subsidence was caused by the densification of embankments and structural backfill materials. Subsidence was usually greatest in the backfill material immediately behind an abutment wall; settlement at one bridge approach on Route 210 was in excess of 0.9 m.[1056] Asphaltic concrete patches were used on approaches along Interstate Routes 5 and 405 to relieve the abrupt changes in profile, but a portion of Interstate Route 210 was closed for an extended period because profile restoration required pavement removal and repaving.[1057]

Roadway damage also included buckled, displaced, and cracked pavement. Youd and Olsen[495] describe this type of damage in areas where there was tectonic rupturing at the surface.

> Tectonic displacements in the San Fernando fault zone [Figure 5 in Volume I, Chapter 3] were responsible for a large part of the pavement damage caused by the earthquake. Wherever structures such as roadways, curbs, and walks crossed zones of tectonic disturbance, they were cracked, buckled, or displaced. In some locations these disruptions blocked the streets; in others, where the magnitude of the tectonic displacement was less or spread over a wider zone, the pavement was cracked and distorted but posed only slight hindrance to traffic.
>
> Thrust faulting was particularly destructive to pavements, leaving them fractured and vertically displaced at all places where fault ruptures were observed to cross them. Compression over finite zones, apparently caused by the thrust faulting, spectacularly buckled rigid pavement sections at vulnerable points...Extensional cracks were common in the zones of tectonic rupture but were generally small enough to cause little hindrance to traffic; however, they will have to be repaired to protect the road base from water intrusion. Similar types of pavement damage were also produced by small gravitational downslope movements outside the zones of tectonic rupture.

Numerous landslides in the San Gabriel Mountains blocked several county roadways, including Big Tujunga Canyon, Little Tujunga Canyon, Lopez Canyon, and Kagel Canyon roads.[1058,1059] Several landslides occurred in cut slopes along Interstate Route 5. The three largest were at highway construction sites, but only one would have blocked traffic. A number of smaller slides, although not blocking roadways, caused pavement sections to heave noticeably.[1057]

Elliott and Nagai[1056] noted that "the San Fernando earthquake, although of moderate size...inflicted greater damage on the California freeway system than any previous earthquake." A cost of $1.64 million was incurred to complete emergency repair work and construction of detours through the damaged interchange areas. Restoration of freeways cost an estimated $12.2 million—$6.5 million for bridge restoration and $5.7 million to restore other facilities.[1057] The estimated cost of the Los Angeles County bridge and highway damage was set at $774,000.[1059]

The Southern Pacific Transportation Company's tracks were damaged in three areas of the San Fernando Valley. In Sylmar, where the main and siding tracks were crossed by the San Fernando fault zone (Figure 5 in Volume I, Chapter 3), the tracks were shifted laterally by 2m, causing kinked rails and one broken rail.[1060] Where the San Fernando Valley Juvenile Hall landslide crossed the tracks, the rail embankment subsided and was laterally displaced more than 0.5 m.[495] The tracks were blocked and damaged by a collapsed bridge structure at the Route 5/210 Interchange (Figure 44). All damaged railroad operating facilities were repaired within 24 hr; cost of repairs was approximately $40,000.[1060]

Although there were 12 airports within a 48-km radius of the epicenter, the only damage to structures was glass breakage at the Hollywood-Burbank and Van Nuys airports. Several aircraft in a hanger were bounced against each other at the San Fernando airport. The most critical problem was the loss of commercial electrical power at several of the facilities. Power outages caused blackouts in terminals and other buildings and prevented the pumping of aircraft fuel from underground tanks. Power in most cases was restored in 8 to 10 hr.[1061]

B. Advances in Antiseismic Lifeline Engineering

As noted by Duke and Moran,[1038,1062] advances in antiseismic lifeline engineering were initiated by the 1933 Long Beach and 1952 Kern County earthquakes and accelerated by the 1971 San Fernando earthquake.* With the exception of nuclear power plants (Section V.E of Volume II, Chapter 3), most of the advances have been limited to lifeline systems in California.[1062] Duke and Moran[1038] describe recent antiseismic design improvements for several lifelines.

Highway Bridges. The California Department of Transportation developed and adopted new procedures for bridge design in 1973. These procedures were adopted by the American Association of State Highway and Transportation Officials (AASHTO) in 1975. However, applications of AASHTO earthquake provisions in new construction have been limited to several western U.S. states. Retrofitting of existing bridges by adding restrainers to interconnect bridge components at hinges has been limited to California.

Telephone Equipment. Some new telephone equipment bracing systems have been improved and some existing equipment bracing has been upgraded, as a result of experience in San Fernando. Control procedures have been developed to minimize breakdowns due to overloading. Major intercity cables use ductile pipe for ducts at active fault crossings and excess slack is provided in the cables, which are laid in sand beds.

Electrical Power Systems. Since about 1933, most California electrical utilities have used earthquake design criteria for their critical facilities which are in excess of those required by local building codes. Some changes in criteria for new electrical equipment have been made since 1971 and considerable research is presently underway. Some existing equipment has been upgraded. Present design criteria for critical equipment are generally based on spectra representing a 50 percent of gravity horizontal ground acceleration combined with a vertical acceleration. System redundancy is recognized as effective in minimizing earthquake effects.

The lateral force design methods for nuclear powered electrical generating plants are at a much higher level of sophistication than those for other electrical generating plants. This is due to the more serious consequences of earthquake induced damage which could cause the release of radioactive materials. The design criteria for nuclear plants are detailed and specific and their enforcement has been stringent.

Pipelines. Damage to underground conduits has been associated with permanent ground displacements. Damage to old lines due to seismic pressure variations and intrusion of foreign objects has been noted. Surveys of underground damage to sewer lines were made by closed-circuit television following the Alaska, 1964, and San Fernando, 1971, earthquakes. Research on earthquake behavior of buried conduits is under way. System redundancy is recognized as a desirable feature.

Piping above grade is generally designed as a structure using a static lateral force of 0.2 gravity or using a spectral approach.

Aqueducts are being planned so that active fault crossings are at or near grade and at right angles to the fault trace. This arrangement should result in damage which can be quickly repaired.

Large natural gas transmission lines have automatic shut-off valves placed at intervals. These valves are actuated by increased flows which can result from pipe ruptures at fault crossings or other locations. Automatic shut-off valves which are actuated by shaking are also used on some customer supply services.

Tanks. Current earthquake design methods for above ground tanks are based on the Structural Engineers Association of California and the Uniform Building Code. Some utilities have adopted more severe standards.

Earth Dams. Beginning with and related to the 1960 Chilean earthquake and the 1964 earthquakes in Alaska and Niigata, Japan, widespread recognition has been given to the strength loss of certain kinds of soils under sustained strong shaking. Over the same time period, the finite element method has been successfully applied to the dynamic design of dams. These two developments have created a means whereby earth dams can be analyzed for earthquake resistance with a much higher order of confidence than in the past. This has provided a foundation for a comprehensive review of the seismic stability of all earth dams in California, with remedial measures required by the Department of Water Resources when appropriate.

The 1971 San Fernando earthquake prompted several other developments to advance the state-of-the-art of lifeline earthquake engineering.

* Several California electrical companies adopted antiseismic design criteria for important facilities in excess of those required by local building codes as a result of the 1933 Long Beach earthquake; the 1952 Kern County earthquake was the catalyst for electric companies improving design criteria for anchoring and bracing equipment and providing flexibility in connected piping systems.[1038]

1. Through the National Oceanic and Atmospheric Administration/Earthquake Engineering Research Institute (NOAA/EERI) Earthquake Investigation Committee,[1044] detailed reports of all lifeline performances were prepared by members of three special subcommittees.[1041-1043] Also, the EERI, through funding from the National Science Foundation, prepared a field guide that describes the current status of lifeline earthquake engineering and a check list to aid investigators in gathering lifeline performance data for future earthquakes.[1038,1063]
2. The American Society of Civil Engineers (ASCE) established an interim Committee on Lifeline Earthquake Engineering (CLEE) in 1973 to define lifeline engineering problems, initiate planning for lifeline research, and develop a charter for the permanent Technical Council on Lifeline Earthquake Engineering (TCLEE).[1039] TCLEE now coordinates and encourages the development of lifeline earthquake engineering and coordinates field investigations of lifeline damage attributable to earthquakes.*[,1038] TCLEE sponsored a major conference on the current state-of-the-art in antiseismic lifeline engineering in August 1977;[1038] the proceedings (37 papers) were published by ASCE.[1064]
3. In California, various government-sponsored groups, including the Joint Committee on Seismic Safety of the California Legislature,[1065] Public Utilities Commission,[1066] and the Los Angeles County Earthquake Commission,[1067] have prepared reports on earthquake-resistant lifeline designs and recommendations for reducing the earthquake hazard to lifelines.[1038]
4. Major California water and electrical utility companies, through the California Water and Power Earthquake Engineering Forum, have coordinated their efforts to develop improved seismic design techniques for new and existing facilities.[1038]

V. ADDITIONAL DAMAGE SURVEYS FOR CONTEMPORARY EARTHQUAKES

It was possible to discuss only five damaging earthquakes in this chapter because of space limitations. Additional examples of building and lifeline damage surveys for destructive earthquakes can be found in the following periodicals: *Bulletin Seismological Society of America, Earthquake Information Bulletin, Earthquake Engineering Research Institute Newsletter, Abstract Journal in Earthquake Engineering, California Geology,* and *Civil Engineering.* Also, the following forums usually have published proceedings that contain papers on damage surveys for destructive earthquakes: *World Conference on Earthquake Engineering, European Conference on Earthquake Engineering,* and the *U.S. National Conference on Earthquake Engineering.*

VI. POST-EARTHQUAKE DAMAGE SURVEYS

The only full-scale test of a structure's ability to resist earthquake forces is a seismic event itself. It is for this reason that post-earthquake engineering investigations are so necessary to develop information for improving the design requirements for earthquake-resistant structures. In the U.S., several organizations, including the National Academy of Engineering, U.S. Geological Survey, National Oceanic and Atmospheric Administration, National Bureau of Standards Institute for Applied Technology, Earthquake Engineering Research Institute, and the American Iron and Steel Institute, have sponsored comprehensive damage investigations for a number of important contemporary earthquakes in various parts of the world.

* TCLEE is comprised of the following committees: Seismic Risk, Gas and Liquid Fuel Lifelines, Transportation Lifelines, Water and Sewage Lifelines, and Electrical Power and Communications Lifelines.[1038]

VII. DYNAMIC ANALYSIS OF EXISTING STRUCTURES AND FOUNDATION MATERIALS

A number of dynamic testing methods are available for determining the dynamic responses of existing structures and foundation materials to prescribed exciting forces.[1068] Data generated from these tests have been used in varying degrees to evaluate and improve earthquake-resistant design principles. Most of the methods are categorized as *forced vibration tests*. This section describes two of the more common forced vibration tests: transient excitations caused by natural earthquakes and explosions and man-excited vibrations—a steady-state sinusoidal excitation. Hudson[1068] describes in detail these and other types of dynamic tests.

A. Transient Excitations by Natural Earthquakes and Explosions

An accurate knowledge of how existing structures and foundation materials respond to the strong vibratory motions of damaging earthquakes is fundamental for the advancement of antiseismic engineering. The damaging levels of ground motion can be obtained only by making actual measurements in the near-field of large earthquakes with strong-motion seismographs or accelerographs[1069] (described in Volume I, Chapter 2). For this reason, accelerographs have been placed in many seismically active regions throughout the world to await damaging earthquakes.[90] Most accelerographs are located in high-rise buildings with a relatively small number sited on other structures such as dams and bridges and various local geologic units. Gates[1070] describes the objectives for instrumenting buildings.

> First, the strong-motion accelerograph would provide the building owner, design engineer, and city safety inspector with a record of the building's response to the earthquake. By analyzing the record, the degree of potential hidden damage in the structure may be determined, thus providing a positive means for evaluating the risks to occupants of the building.
>
> Second, the recorded motions of a building would provide valuable data for review of earthquake design procedures and evaluation of the actual safety factors in the minimum design requirements specified by the building code. The results of such reviews and evaluations hopefully would bring about improved design procedures for earthquake-resistant buildings and greater assurance of structural safety through improved minimum design requirements.
>
> There are several secondary objectives behind strong-motion instrumentation. One is to study the influence of site conditions on the ground motions delivered to the building. Do certain sites present a greater earthquake hazard than others? What building types are particularly vulnerable under given site conditions? By instrumenting many types of buildings on various sites with different geological characteristics, it is possible that the answer to these questions may be determined.
>
> Another secondary objective is improved analytical procedures and mathematical modeling techniques. By applying the ground motions recorded at the base of a building as a forcing function to a mathematical model of the building, the engineer can create the motions, forces, and stresses in the building by sophisticated computer analysis techniques. The accuracy of this solution is measured by the closeness of fit between recorded and calculated acceleration response in the upper stories of the structure. The mathematical model is a primary variable affecting the accuracy of the solution. Thus, strong-motion records can provide valuable information on the credibility of mathematical models and analytical procedures.

The optimum condition for determining the influence of foundation materials on ground motion is to emplace accelerographs at *free-field* or *fully external sites* (i.e., away from structures). This is because the recorded base motion of a structure can differ from that which would have been recorded had there been no structure at the site because of possible structure-foundation interaction.[1071]

The first strong-motion instruments measured ground accelerations as a function of time in the three principal directions of motion—longitudinal, transverse, and vertical; the graphical records are called accelerograms (Figure 35 in Volume I, Chapter 2). In the late 1960s, an extensive program of accelerograph data processing was initiated at the Earthquake Engineering Research Laboratory, California Institute of Technology. Techniques were developed to reduce accelerograph analog responses to digital form and perform corrections for long-period distortions and high-frequency errors.[1072-1076] In addition, mathematical integration procedures were perfected that now make it possible to process the *acceleration history* to produce accurate *velocity and displacement histories* of the ground motion as a function of time by integrating the accelerogram once and twice, respectively (Figure 45).[1077-1081] Because of the interest in long-period motions for engineering applications and the complexities involved with numerical integration, a new type of accelerograph was introduced in the early 1970s with three long-period transducer elements for measuring the amplitude of ground displacement vs. time for certain frequency ranges.[1069,1082]

The most complete characterization of dynamic ground vibrations is a specification of the motion "in three independent spatial coordinates for every instant of time in terms of acceleration, velocity, or displacement."[513] With reference to the damage potential of ground motion, the following are important factors:

Peak values of acceleration, velocity, and displacement—Each generally increases with earthquake magnitude and decreases with increasing distance from the area of energy release for an earthquake of any magnitude; damage generally tends to increase directly with the amplitude of ground motion.[513,1083] Of the three ground motion parameters, the maximum horizontal acceleration is usually the parameter specified for earthquake-resistant design criteria.[1084]

Mean frequency of the dominant pulses in the three histories—Structures and in some cases surficial deposits may respond in resonance if their natural frequencies approximate those of the ground motion. Large deformations and stresses can occur in a structure or surficial deposit if the ground vibrations include significant amounts of energy at frequencies close to the natural frequencies of the host system.[513,1083]

Duration of the most intense portion in each history—Duration influences damage because the failure mechanism in structures and surficial deposits is dependent upon the number of induced stress cycles; hence, there is a direct relationship between duration and the potential for damage.[513] One duration measure frequently used in earthquake engineering is the time during which the acceleration equals or surpasses an amplitude threshold of 0.05×g.[1085]

Strong-motion data can also be used as input to produce *Fourier amplitude spectra of acceleration* and *acceleration, velocity, or displacement response spectra* to further define ground motion. Both types of spectra are calculated by analog or high-speed digital computers.[119] A Fourier spectrum is a method for displaying the frequency content of an accelerogram. Motion is perceived as being comprised of an infinite number of harmonic waves with different amplitudes and frequencies. In diagrametric form, the amplitudes of the harmonic components are plotted as a function of their frequency. Peaks on the resulting spectrum curve represent frequencies at which there were relatively large inputs of energy into the host system.[501,1083,1086] The north-south accelerogram and Fourier amplitude spectrum from the May 17, 1976 Gazli, Soviet Union earthquake (M_s = 7.2) are shown in Figure 46.

A response spectrum represents the peak acceleration, velocity, or displacement response for a series of single-degree-of-freedom systems having viscous damping to a prescribed input ground motion. The response for structures in their elastic range is a function of damping and natural frequency, or its inverse, the natural period. A plot of peak response as a function of acceleration, velocity, or displacement for systems of identical damping represents a response spectrum for the parameters considered. A family of such curves for various levels of damping represents a response spectra. A high peak in a response curve represents a high response (deformation) level and a

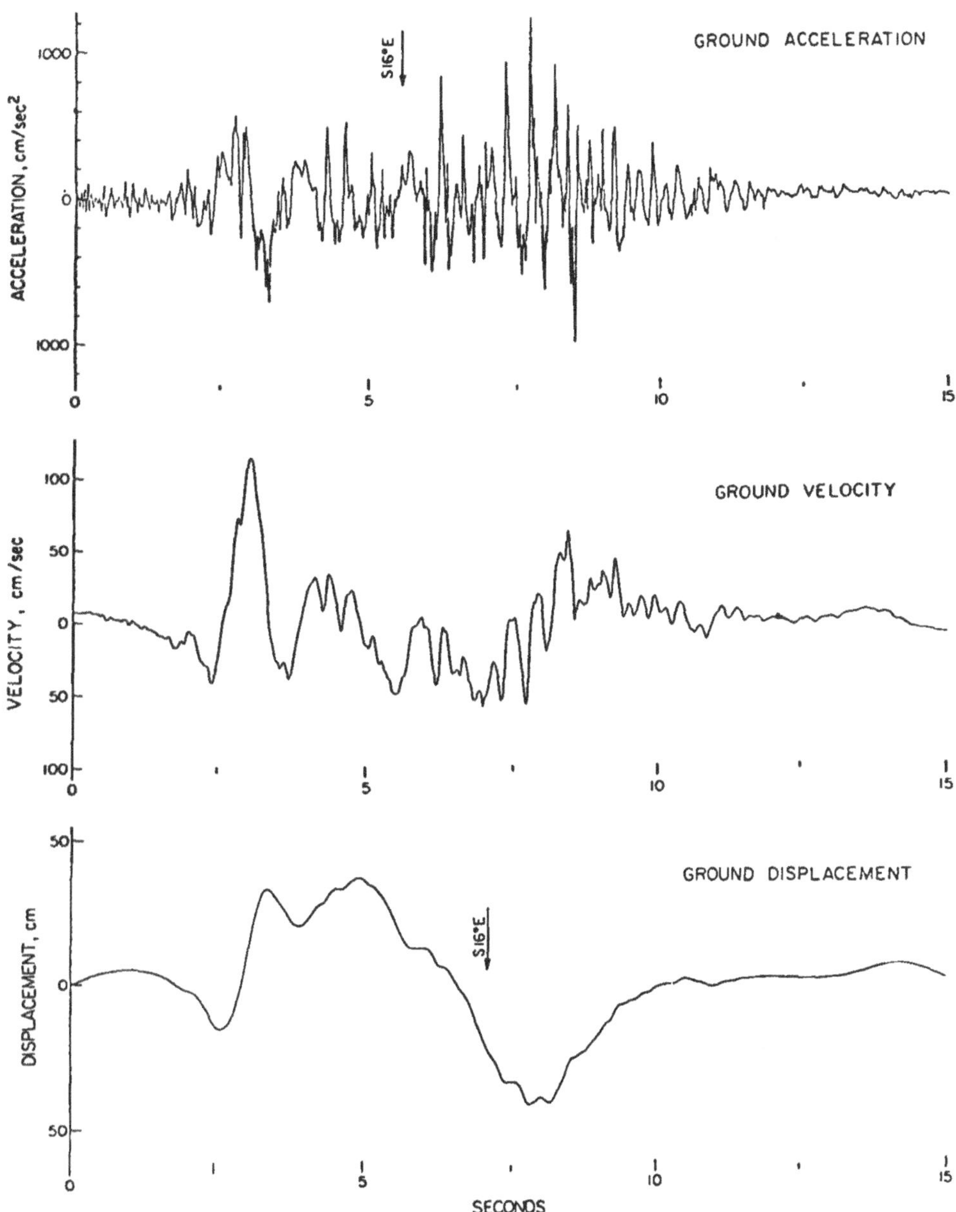

FIGURE 45. *Acceleration, velocity, and displacement histories* of the S16°E component of horizontal ground motion at the Pacoima Dam accelerograph site for the February 9, 1971 San Fernando, California earthquake (M_L = 6.4). Note that the maximum acceleration was 1.25×g, nearly double that recorded for any previous earthquake. The accelerograph site was 8 km from the epicenter and 4 km from the nearest surface faulting. It should be noted that the rupture may have propagated southward and upward beneath the dam. The instrument was emplaced on a gneissic granite-diorite ridge adjacent to the dam. (From Trifunac, M. D. and Hudson, D. E., San Fernando, California, Earthquake of February 9, 1971, Vol. III, Benfer, N. A., Coffman, J. L., and Bernick, J. R., Eds., U.S. Government Printing Office, Washington, D.C., 1973, 386.)

high concentration in the motion of waves of the frequency or period corresponding to the peak.[932,1083,1087] Figure 47 displays the S80°E accelerogram for the March 22, 1957 San Francisco earthquake (M_L = 5.3) recorded at Golden Gate Park and the acceleration response spectra for various degrees of damping. Note that the undamped

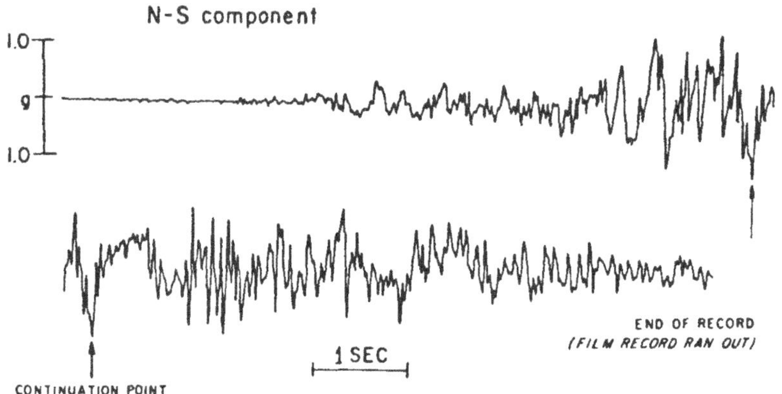

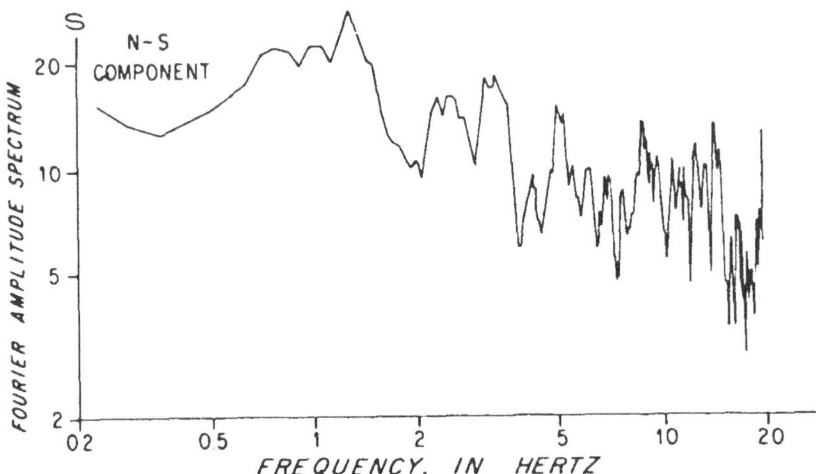

FIGURE 46. North-south *accelerogram* and *Fourier amplitude spectrum* from the May 17, 1976 Gazli, Soviet Union earthquake (M_s = 7.2). The accelerograph was located at the Karakyn Point seismic station, approximately 10 km from the epicenter. Note that the maximum energy input occurred at a frequency of approximately 1.5 Hz. (From Pletnev, K. G., Shebalin, N. V., Shteinberg, V. V., and Rojahn, C., Seismic Engineering Program Report, January—April 1977, U.S. Geological Survey Circular 762-A, 1977, 26, 28.)

response (top curve) is approximately 0.75 × g for an oscillator with a period of 0.25 sec, 0.10 × g for a 1-sec period, and 0.03 × g for a 2-sec period. For 20% damping (bottom curve), the response is approximately 0.20 × g for a period of 0.25 sec and about 0.015 × g for a 2-sec period.[932] As noted by Degenkolb,[932] the acceleration spectra in Figure 47 indicate that tall buildings, with longer natural periods, would have smaller inertia forces exerted on them than stiff, short-period buildings. Fourier and response spectra techniques are described in more detail by Clough,[929] Degenkolb,[932] Housner,[501,1086] Hays,[1087] and Hudson.[1088,1089]

The U.S. currently has the largest number of operational strong-motion networks and several are described in the following paragraphs. The national strong-motion program was initiated by the Coast and Geodetic Survey in 1932 with the emplacement of nine instruments in structures in California. By 1962, the network consisted of only

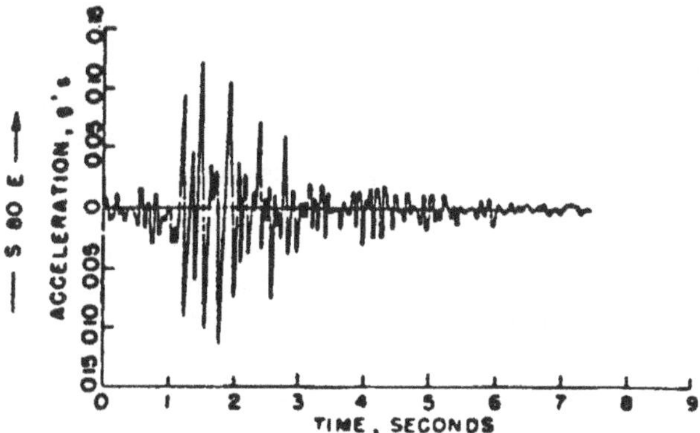

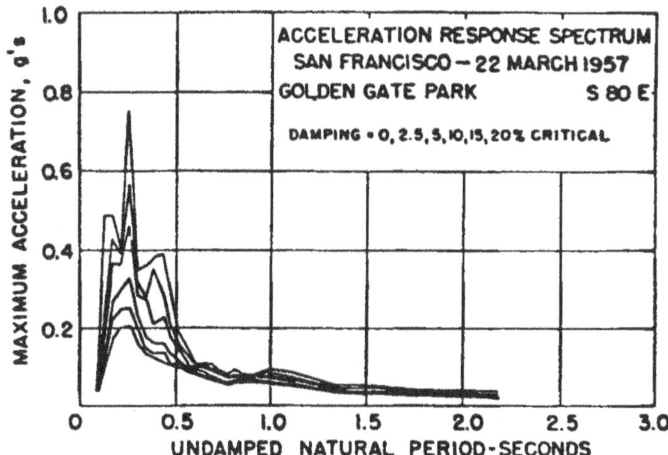

FIGURE 47. S80°E *accelerogram* and *acceleration response spectra* for the March 22, 1957 San Francisco earthquake (M_L = 5.3). For the response spectra, the top curve represents 0% damping and the bottom curve 20% damping. (From Degenkolb, H. J., *Earthquake Forces on Tall Structures,* Booklet 2717A, Bethlehem Steel Corp., Bethlehem, Pa., 1977, 5. With permission.)

70 accelerographs. However, improved and less expensive commercial strong-motion seismographs were introduced in the mid-1960s, enabling the network to undergo rapid expansion. The network is now comprised of more than 850 accelerographs in 36 states, Puerto Rico, and Central and South America; most of the instruments are in the western states. Since 1973, the national network has been managed by the Seismic Engineering Branch of the Office of Earthquake Studies, U.S. Geological Survey and supported by the National Science Foundation in cooperation with numerous governmental and private agencies and organizations.[1082,1090,1091] The U.S. Geological Survey publishes the *Seismic Engineering Program Report* * three times yearly to apprise the ever-growing community of strong-motion data users of the availability of data recovered by the national network and abstracts of recent strong-motion reports.[1090]

* Free upon application to Branch of Distribution, U.S. Geological Survey, 1200 South Eads Street, Arlington, Va. 22202.

Until recently, the emphasis of the national strong-motion program was on measuring motions in or near buildings. However, the program now includes free-field regional arrays to measure the dynamic responses of various types of surficial geology to seismic vibrations. Two of the first regional sites were the Puget Sound area of Washington[1092,1093] and the San Francisco Bay region.[1094] The latter site has been instrumented with a linear array of accelerographs, accelerometers,* and seismoscopes** to (1) simultaneously measure ground motions on bedrock, alluvial materials, Bay mud, and hillside materials for nearby and more distant earthquakes, (2) determine the attenuation of ground motion near the San Andreas and Hayward faults, and (3) determine the variation of ground motion on opposite sides of the San Andreas and Hayward faults. The APEEL (Andreas-Peninsula Earthquake Engineering Laboratory) array is comprised of 14 free-field stations, where the surficial geology is well known, and extends between and across the San Andreas and Hayward faults (Figure 48). Accelerometers at one site are emplaced in a bore-hole drilled through approximately 200 m of Bay mud and silty sand to the top of Franciscan bedrock (Figure 48). Current plans call for additional arrays to be established in Alaska, California, the Yellowstone Park region, and the Mississippi embayment.[1095] A major advantage of accelerograph regional arrays is that earthquake source and travel path factors affecting ground motion (e.g., stress drop, source dimension, radiation pattern, nature and configuration of geologic structure along the path of propagation, and distance from the energy source) tend to cancel for the same event and the effects of differences in surficial geology, including topography, become paramount.

There are several smaller strong-motion programs at the federal level. For example, a number of dams have been instrumented with accelerographs by the Army Corps of Engineers and the Bureau of Reclamation,[1082] and the Nuclear Regulatory Commission requires the installation of strong-motion seismographs at all nuclear power plants (Appendix F).[973]

The California State network, the second largest in the U.S., was established by Senate Bill (SB) 1347 in 1971 and is managed by the Strong-Motion Instrumentation Program (SMIP) of the California Division of Mines and Geology. To fund the program, SB 1347 requires all counties and cities to collect a fee from applicants for building permits; the fee amounts to a few cents per $1000 valuation on new construction.[1096,1097] Rojahn[1098,1099] reports that on the basis of recommendations made by a special ad hoc committee, 21 regions will be instrumented under the *building instrumentation phase* of the California Strong-Motion Instrumentation Program. The regions were selected on the basis of population density, location of buldings already instrumented, and the probability for damaging earthquakes. Strong-motion seismographs are to be placed in a variety of buildings, but the instrumentation of low-rise structures will be emphasized.[1098,1099] As described by Wootton,[1100] the *subsurface phase* of the program was implemented recently to record the response characteristics of subsurface materials to earthquake motions. At each station, two bore-holes are drilled, one into bedrock and the second to a depth halfway between the surface and the bedrock. A three-component (triaxial) accelerometer is placed in each hole, and another triaxial accelerometer is placed at the ground surface. All three instruments are attached to a single data recorder. By April 1977, two stations were in operation, and three more were in the planning stages. Wootton[1100] acknowledges that the under-

* An accelerometer measures horizontal and vertical ground motion with no *in situ* recording capability. The response data are telemetered to a recorder at some accessible location.

** Seismoscopes, described in Volume I, Chapter 2, measure horizontal ground motion in the near-field of an earthquake by the angular deflections of a pendulum (Figure 36 in Volume I, Chapter 2). The deflections are recorded on a smoke glass plate by a metal stylus (Figure 37 in Volume I, Chapter 2).

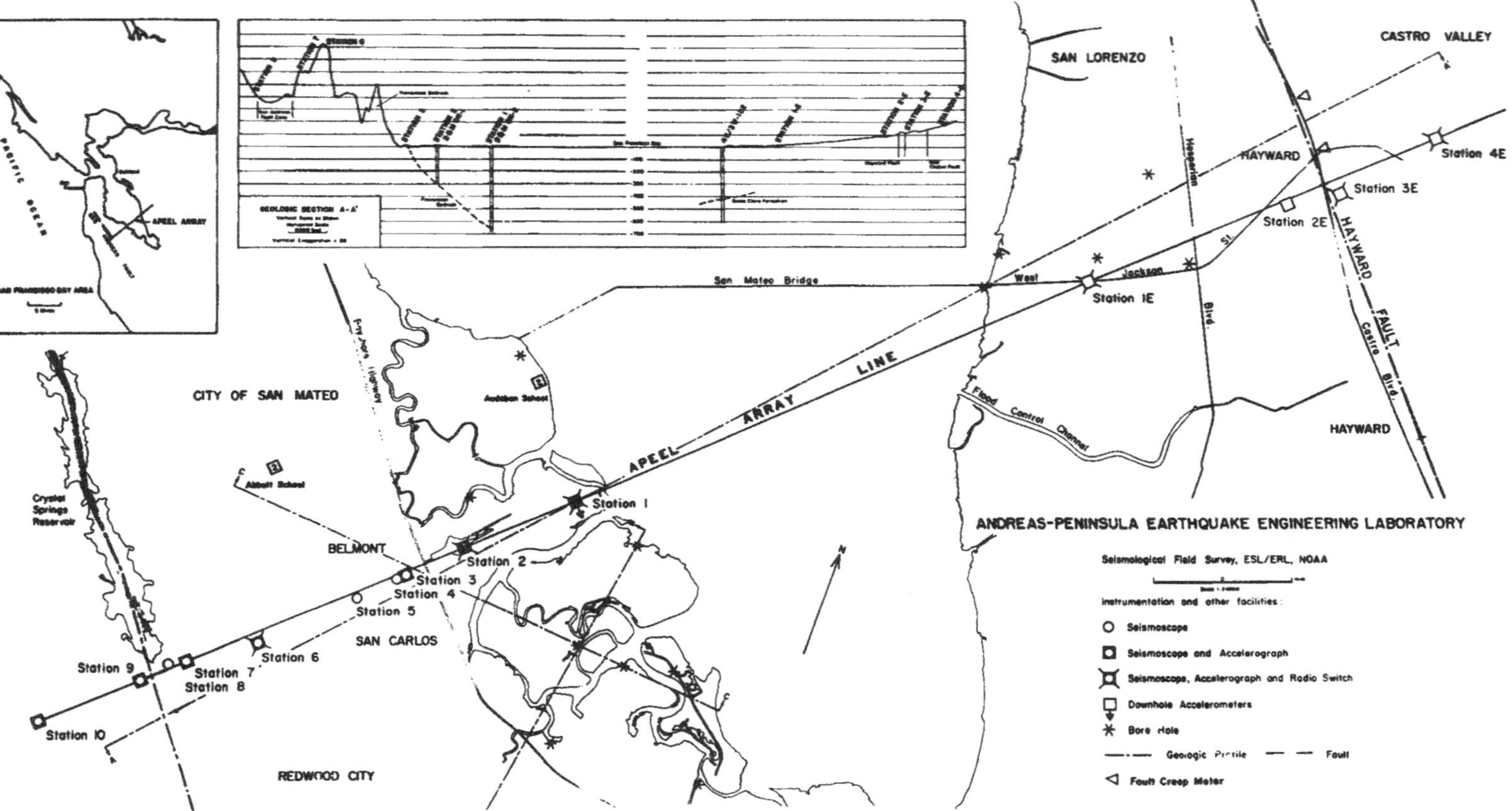

FIGURE 48. The APEEL (Andreas-Peninsula Earthquake Engineering Laboratory) strong-motion test site in the San Francisco region. (From Morrill, B. J., The APEEL Array: A site Study, NOAA Technical Report ERL 245-ESL 22, U.S. Government Printing Office, Washington, D.C., 1972, 7.)

FIGURE 49. *Horizontal traction* or *plate and hammer method* used for identifying subsurface layers of rock or soil plus their configurations and elastic properties. Mechanics of the method are described in the text. (Courtesy of Jay H. Power, California Division of Mines and Geology.)

ground responses are important for two basic reasons: "(1) direct application to the design of deeply founded and underground structures, and (2) further understanding of how motion responses vary as earthquake energy is propagated upward from the underlying bedrock to the surface." The long-range plan for the state program calls for the deployment of 1864 instruments over a 60-year period at a cost of $12 million. The completed network will consist of 520 instruments at free-field rock and soil sites, 400 instruments in buildings, and 944 instruments on dams, bridges, and various utility structures.[1101]

In order to gain an understanding of the influence of a site's geology in affecting ground motion, Power and Real[1102] report that site investigations for accelerograph stations in the California network include field measurements of horizontal shear wave velocities to identify subsurface layers of rock or soil plus their configurations and elastic properties. This is accomplsihed by the *horizontal traction* or *plate and hammer method* (Figure 49) that was developed by the Earthquake Research Institute, Tokyo University. The mechanics of the method are as follows:

1. A large horizontal plank, held firmly against the ground by the front wheels of a truck, is hit on one end with a large wooden hammer (Figure 49).
2. The horizontal hammer blow imparts a traction or shearing stress tangential to the surface of the ground, thereby producing a ground motion rich in horizontal shear energy.
3. The waves are detected by geophones emplaced at increasing distances from the energy source along an axis parallel to the plank.
4. By plotting the first-wave traveltime to each sensor vs. source distance, distinct seismic layers are identified that normally correspond to different geological units.[1102]

The City of Los Angeles enacted an ordinance on July 1, 1965 that requires accelerograph instrumentation in (1) every new building over six stories in height with a floor area of 5576 m^2 or more and (2) all new buildings over ten stories high, regardless of floor area. The buildings must be instrumented with three approved accelerographs at unobstructed locations in the basement, midstory, and near the top of the structure.*[938,1082] A similar provision was later added as an appendix to the Uniform Building Code for Seismic Zone 3 (now Seismic Zones 3 and 4—Figures 7 and 8 in Volume II, Chapter 3). The appendix has been adopted by more than 50 cities, mostly in California.[937,1082] The Joint Building Committee of the Structural Engineers Association of Northern California and the San Francisco Section of the American Society of Civil Engineers believes that the Uniform Building Code approach could be better served if buildings were selected on an engineering basis, enabling many types of buildings, not just high-rises, to be instrumented.[1103]

Two strong-motion programs were established as a direct result of the catastrophic damage incurred by hospital buildings (Figure 16 in Volume II, Chapter 3) and bridges (Figure 44) during the February 9, 1971 San Fernando earthquake. First, the Veterans Administration has installed accelerographs in 65 of its hopsital buildings in Seismic Zones 2, 3, and 4 (Figures 7 and 8 in Volume II, Chapter 3); the strong-motion data will be used to make revisions in the building standards for VA facilities in earthquake-prone areas.[965,1104] Second, the National Oceanic and Atmospheric Administration officially began a bridge instrumentation program in December 1971 when two accelerographs were installed on a bridge at the Interstate 5/14 Interchange north of San Fernando.** The initial phase of the program is centered on Interstate Route 5, but the network will eventually cover the entire California freeway system.[1105] The Earthquake Engineering Section of the Bridge Department, California Division of Highways, is in the process of installing strong-motion instruments on bridges in the Los Angeles and San Francisco regions.[1106] The importance of instrumenting bridges in California has been summarized by a NOAA spokesman:[1105]

> Highways have always been considered prime candidates for strong-motion measurements, but not particularly cost-effective ones. Early instruments were too large and too costly to be used efficiently along highways, and, besides, roads seemed less critical than structures containing life. . . .
>
> But highways have changed, especially in California. Now ribbons of concrete and steel, immense cantilevered spans and banked roadways curve hundreds of feet into the air, and represent as complicated a set of structures as man is likely to build. Moreover, the car-based culture of California makes this intricately woven net of overpasses, interchanges, and multilane freeways a major conduit of life. Until the San Fernando experience, relatively little was known about how these elaborate traffic-bearing systems would respond to earthquake induced forces; it has become clear that much more must be measured to obtain a full picture of that response.

Because large earthquakes are fortunately rare and because of the slow growth of accelerograph networks, only a limited number of useful accelerograms for large earthquakes have been obtained for even the most seismically active regions of the world. For example, prior to the 1971 San Fernando earthquake, there were only about ten accelerograms from American earthquakes that were useful for antiseismic engineering design purposes.[1096] Although the San Fernando earthquake triggered 241 accelerographs between 8 and 370 km from the epicenter—the largest number yet activated for a single event—90% of the instruments were located in downtown Los Angeles where damage was minor and fewer than 20 of the instruments were on free-field sites.[1107] Little or no strong-motion data are yet available for areas of low seismicity, and historical intensity data often must be used to assess the severity of ground mo-

* The vertical arrangement enables the basement accelerograph to measure input ground motions and the midheight and upper-level accelerographs to measure structural responses.[1068] Motion amplitudes usually increase with increases in height in the same building.

** Since 1973, the program has been carried out by the U.S. Geological Survey.

tion.[1108] For example, it was not until June 13, 1975 that an earthquake in the New Madrid seismic zone* (m_b = 4.0 to 4.25) produced accelerograms; in addition, the accelerations for this earthquake were the largest ever recorded in eastern North America.[1109]

Because of the lack of suitable strong-motion data, small earthquakes and underground nuclear explosions** at the Nevada Test Site (NTS) have been used as sources of transient motion for dynamic test excitations.[90,535,1110-1114] At NTS, special buildings are constructed near the test areas to observe damage patterns. According to Bolt,[90] the results to date indicate that the maximum velocity of ground motion is an effective indicator of damage. For example, plaster cracks in new construction do not become significant until ground velocities exceed 20 cm/sec, and structural damage to wood-frame structures does not occur until velocities reach 150 cm/sec. Strong-motion seismographs also are used at NTS to measure ground shaking at various distances from ground zero and on different types of surficial geology.[90] In addition, Blume[1110,1112,1113] reports that as a byproduct of the Nuclear Regulatory Commission's underground nuclear safety program, new data are being generated on the dynamic responses of many high-rise buildings in Las Vegas to ground motions resulting from underground nuclear explosions at NTS. Much of the knowledge gained from these tests is applicable to the problem of how high-rise buildings respond to ground motion resulting from natural earthquakes and determining what the dynamic characteristics of tall buildings actually are.[112]

Recent studies using nuclear explosions at NTS and small earthquakes for determining the effects of local surficial geology on ground motion have been completed for the San Francisco Bay region.[535,1111,1114] Measurement of ground motion at up to 99 locations indicate that local geological conditions can change dramatically the characteristics of ground motion. In qualitative terms, the response data indicate that the effects of amplified ground motion are least for bedrock sites, intermediate for alluvium sites, and greatest for sites underlain by artificial fill and Bay mud.[535]

B. Man-Excited Vibrations

A relatively new procedure for determining a structure's response to an external excitation is being performed by the Special Projects Party, U.S. Geological Survey. Personnel use the *man-excited vibration technique* to ascertain how a particular building responds to a small external force, how quickly it damps out the motion, and how it might respond to a much larger external force, such as a natural earthquake. The external force is applied by a person (or persons) shifting his weight synchronously with a building's natural vibrational period (Figure 50). Even though a tall structure may move only a few hundredths of a centimeter, it is an ample response for measurement. Up to 12 seismometers can be placed in various levels of a building, and the vibration spectra are recorded on a 12-channel oscillograph. The man-excited method can determine the first, second, and third vibrational modes and torque motions of a building. A strain-gauge platform is now used by the weight shifter, enabling the input forces to be measured simultaneously with structural responses. Kenneth King, Chief of the Special Projects Party, was primarily responsible for developing the man-excited vibration technique into a fully operational procedure.[1115]

One goal for this technique is to pinpoint earthquake risk. For example, the testing

* The New Madrid seismic zone, centered in southeast Missouri, was the site of a series of large earthquakes in 1811-1812.

** As noted by Bolt,[90] accelerogram interpretations for nuclear explosions must be cautionary because of the shallow and restricted nature of the energy source compared with fault rupture at depth for natural earthquakes.

FIGURE 50. *Man-excited vibration testing* of Caesar's Palace, Las Vegas, Nevada. Before completion, the building (top left) was instrumented with seismometers (11th-floor device, top right). A man is shown shaking the building in its minor axis (middle row) and its longitudinal axis (bottom row). (From **Anon.**, *Earthquake Inf. Bull.*, 4, 8, 1972.)

may reveal discontinuities in the vibrational response of a structure, suggesting areas of potential structural weakness that could fail during future seismic shaking. For buildings that had been pre-earthquake tested, post-earthquake tests might locate areas in a building where damage had occurred, but which is not readily apparent to an engineering inspection team.[115]

This technique is replacing the traditional *mechanical shaker*[1116] for exciting real buildings, especially those where the construction has been completed. Most shakers use a rotating eccentric weight for producing the input force. They have a distinct disadvantage over the man-excited technique because the installation, operation, and demounting of the shaker can interfere with many kinds of activity that take place within a completed building.[1115]

Index

INDEX

A

B

C

K

L

M

N

O

P

R

S

T

U

V

W

Printed and bound by CPI Group (UK) Ltd, Croydon, CR0 4YY
22/10/2024
01777632-0020